Telecoms in the Internet Age

Telecoms in the Internet Age

From Boom to Bust to . . . ?

MARTIN FRANSMAN

OXFORD
UNIVERSITY PRESS

OXFORD

UNIVERSITY PRESS

Great Clarendon Street, Oxford OX2 6DP

Oxford University Press is a department of the University of Oxford.
It furthers the University's objective of excellence in research, scholarship,
and education by publishing worldwide in

Oxford New York

Auckland Bangkok Buenos Aires Cape Town Chennai
Dar es Salaam Delhi Hong Kong Istanbul Karachi Kolkata
Kuala Lumpur Madrid Melbourne Mexico City Mumbai Nairobi
São Paulo Shanghai Singapore Taipei Tokyo Toronto

with an associated company in Berlin

Oxford is a registered trade mark of Oxford University Press
in the UK and in certain other countries

Published in the United States
by Oxford University Press Inc., New York

© Martin Fransman, 2002

British Library Cataloguing in Publication Data

Data available

Library of Congress Cataloging in Publication Data

Data available

ISBN 0–19–925700–0

10 9 8 7 6 5 4 3 2 1

Typeset by Newgen Imaging Systems (P) Ltd, Chennai, India
Printed in Great Britain
on acid-free paper by
Biddles Ltd, Guildford and King's Lynn

PREFACE

My interest in the Telecommunications (Telecoms) Industry began in the mid-1980s when I had my first long stay in Japan. I had gone to Japan to try and understand better that country's remarkable economic catch-up process, particularly in the information and communications industries. I soon discovered, to my initial surprise, that these industries and their associated companies—now household names like NEC, Fujitsu, Hitachi, and Toshiba—had their roots in the nineteenth-century Japanese telecoms industry (as elaborated upon in my *Japan's Computer and Communications Industry*, Oxford University Press, 1995).

In the rest of the world too, the Telecoms Industry has been midwife to many of the key industries that have been the mainspring of economic growth in the post-Second World War global economy. To take one example, the transistor, invented in 1948 in AT&T's Bell Laboratories,[1] was to provide a major stimulus to the computer, consumer electronics, and semiconductor industries that in turn created the conditions for the growth of complementary industries such as software. Japan, and later countries like Korea and Taiwan, were able, for many complex reasons, to take advantage of the new opportunities for profit and economic growth opened up by these industries.

The mid-1980s was also a particularly important time in the evolution of the Telecoms Industry. At around this time, though for very different reasons, Japan, the UK, and US decided to liberalize their telecoms industries and allow new competitors to compete with their hitherto state-controlled monopolies, NTT, BT, and AT&T. These companies *were* largely the telecoms industries in these countries. How would they cope with the radical winds of change? How would the Telecoms Industry evolve, both in these countries and globally? These were the kinds of questions I was forced to address as my research subject was transformed by these events. The continuing story of the fortunes and failures of the Big Five incumbent network operators—AT&T, BT, Deutsche

---

[1] For more detail see Fransman (1995: 45–6).

Telekom, France Telecom, and NTT—constitutes one of the main themes that threads its way through this book, as do the stories of the vigorous new entrants that emerged to challenge them.

In the mid-1990s, the Internet first emerged as a phenomenon affecting masses of ordinary people, surprising even its creators like Vinton Cerf, to some the father of the Internet, and key players such as Bill Gates and his company Microsoft. Here too the Telecoms Industry was a midwife, facilitating from the early-1960s the communications between remotely located computers that made interactive computing possible and (as shown in Chapter 2) launching some of the processes that would lead to the Internet as we know it today.[2]

But although the Telecoms Industry was a midwife to the Internet, it was also transformed by its creation. The problem was that it was not clear precisely how the Telecoms Industry was changing under the influence of the Internet. What were the dynamics driving this changing industry in a process of meta-morphosis? Were new industrial structures and processes emerging to replace the old Telecoms Industry and, if so, what were they and how did they work? The answers to these questions constitute another major theme addressed throughout this book, most explicitly in Chapter 2.

The late-1990s saw the emergence of a roller-coaster ride that swept up not only the entire Telecoms Industry but also the so-called 'tech companies' (including the now notorious dotcoms) and a significant part of the global financial markets. The Telecoms Boom from around 1996 carried telecoms companies to new heights through the creation of unprecedented stock market values. However, from March 2000, boom turned to bust and by September 2001 about $2,500 billion worth of market value was destroyed. This raised further questions. Why did the boom and bust occur? How did this financial phenomenon relate to the dynamics and evolution of the changing Telecoms Industry? Could anyone be blamed for the boom and bust? These questions are answered in Chapter 1.

The questions I was asking also raised further questions regarding an appropriate analytical framework for providing answers. Most of my fellow economists involved in telecoms were applying their paradigm (based largely on rational choice and optimization) to the derivation of regulatory rules for the industry. In areas such as interconnection policy, these rules would be used to harness the forces of competition in order to produce optimal outcomes. Since there is a specialized and voluminous literature on the economics of regulation as applied to telecoms, and since it is not these areas of regulation that are my prime concern here, little reference is made to this important body of literature in the present book.

---

[2] As shown in Chapter 2, however, the Telecoms Industry was a reluctant midwife, with many telecoms engineers and researchers in the industry strongly believing that packet switching, a key Internet-related innovation that came from the computer field, was 'stupid'.

However, this orthodox economics cognitive framework was not suitable to tackle the questions I was asking, bedevilled as they were by uncertainty, sheer complexity, and at times mistaken beliefs. As shown in many places in this book, the world in which the decision-makers in the changing Telecoms Industry had to make their decisions often seemed so unclear as to make rationally chosen optimal choices seem like part of cloud cuckoo land. Cognitive frameworks they certainly developed in order to guide their way, and indeed they had no option but to develop them if they wanted to make any decisions at all in their uncertain and unclear world. And these cognitive frameworks embodied their understanding of the industrial structures, institutions, and processes of change that they believed would influence the outcome of the decisions they made.

In this book, a further important theme is the analysis of these cognitive frameworks—often referred to as visions—and the roles they have played in shaping the evolution of the Telecoms Industry in the Internet Age and its ups and downs. But these cognitive frameworks are not treated as the constructs of competent calculators, designed to produce optimal outcomes. Rather, they are treated as more hesitant conjectures that the creator is only too-well-aware may turn out to be incorrect. And as the evolution of the Telecoms Industry traced in this book shows, all too often they were incorrect.

More generally, the cognitive framework that has implicitly informed the present book may be described as institutional evolutionary economics of the kind pioneered by Dick Nelson and Sidney Winter in their book, *An Evolutionary Theory of Economic Change*, and elsewhere and developed by many other authors since. However, as already made clear, the framework used here is an approach which insists that ignorance and mistakes are more prevalent than is usually acknowledged and as a result that human beliefs, though crucial for human action, often rest on foundations that are less secure than is usually admitted. The beliefs held during the Telecoms Boom and Bust analysed in Chapter 1, or at various critical junctures in the evolution of the Telecoms Industry as shown in Chapter 2 and elsewhere in this book, provide rather dramatic examples.[3]

In attempting to understand the evolving Telecoms Industry in the Internet Age, I have had the good fortune to receive the help of many people and institutions. One of my greatest debts is to my Japanese friends and colleagues who first opened my eyes to many of the issues raised in this book and continue to provide rich insights. I am particularly grateful to many in NTT, the Japanese incumbent network operator that has been at the heart of many of the processes analysed here. There are too many friends and acquaintances from NTT who have helped me to include them all here. However, mention must be made

---

[3] As the reader will have gathered, I have, therefore, not been inclined to follow the advice of an anonymous economist reviewer whose recommendation was for me 'to develop a detailed dynamic model of the operation of the [Telecoms] industry in order to determine the long-run equilibrium outcome'. I hope that I have said enough to indicate why I feel this approach would not be particularly illuminating and would not answer the questions I am posing.

of Mr Jun-Ichiro Miyazu, President and CEO of NTT, Mr Norio Wada, President-elect of NTT, Mr Kanji Koide, Senior Vice President of NTT in charge of corporate strategy, Dr Shigehiko Suzuki, Senior Vice President of NTT in charge of R&D, Mr Michio Takeuchi, Senior Executive Vice President of NTT West, Dr Keiji Tachikawa, President and CEO of NTT DoCoMo, Mr Kiyoyuki Tsujimura, Senior Vice President of NTT DoCoMo, and Dr Toshiharu Aoki, President and CEO of NTT Data. Friends of long standing, who have also over the years provided important insights, include from NEC, Dr Hisashi Kaneko, former CEO, and from Fujitsu, Mr Shigeru Sato, President of Fujitsu Laboratories, Dr Iwao Toda, and Dr Yukou Mochida. Professor Shumpei Kumon and his colleagues at GLOCOM in Tokyo could always be relied upon for challenging thought and stimulation. Mr Teruaki Ohara, CEO of InfoCom Research in Tokyo, and his colleagues Mr Kiyoshi Fujita, Mr Takahiro Ozawa, Mr Nobuki Hori, Ms Keiko Hatta, and Ms Kunie Yokoyama also provided important insights and assistance.

In the early 1990s, I had the good fortune to meet Dr Arno Penzias, then Vice President Research, AT&T Bell Laboratories. The insights that he and his colleagues in Bell Labs gave me into the changing world not only of the Telecoms Industry but more generally the Infocommunications Industry have left their mark on many of the discussions here. Similarly my views have been influenced by the large numbers of interviews I have been fortunate to have with leaders in the Telecoms Industry. Although again too many to record here, mention must be made of Sir Peter Bonfield, then CEO of BT, Mr Paul Chisholm, then CEO of COLT, Mr Mike Grabiner, then CEO of Energis, and the board members and senior executives of many telecoms companies including Deutsche Telekom, Ericsson, France Telecom, Lucent, Nokia, Nortel, Siemens, Sonera, Telecom Italia, Telia, Viatel, Vodafone, and WorldCom.

I am also grateful for the discussions and stimulation I received from friends when I held visiting posts at RCAST, Tokyo University in Japan, Chalmers University in Sweden, IDEFI associated with the University of Nice in France, and ICER associated with the University of Torino in Italy. I must also acknowledge financial support from several sources including JSPS in Japan, several European Union TSER Programmes, the ESRC Innovation Programme, and JETS.

It has, as always, been a joy to share ideas, work with and learn from close friends. Particularly important has been Sadahiko Kano, former Senior Vice President of NTT who has spent several years as visiting professor at the Institute for Japanese-European Technology Studies (JETS) at Edinburgh University. Sadahiko has long been an extremely thoughtful analyst of the global telecoms industry as well as a global participant shaping part of its evolution. Jon Sigurdson, Cristiano Antonelli, Eric Bohlin, Jackie Krafft, and Klaus Rathe also deserve special mention. I must also include the academic work of Brian Loasby that seldom fails to inspire me for reasons that will be obvious to anyone reading this book who is also familiar with his work.

My colleagues in JETS—Lynne Dyer, whose friendship and competence has been a pillar of support, and Tim Bolt and Ian Duff, whose excellent research assistance has been indispensable—were crucial for the present work. Shoko Tanaka over many years has provided invaluable friendship and support. Finally, I must acknowledge with particular gratitude the support of my immediate family—Tammy, Judy, Karrie, Jonathan, Dee, Denny, and Terry— who have had to cope with my booms but also, unfortunately, my busts. This book is dedicated to the memory of my father, Elie, who taught me more than he ever would have realized.

This Book is Dedicated to
The Memory of My Father, Elie

CONTENTS

LIST OF TABLES

LIST OF FIGURES

ACRONYMS AND ABBREVIATIONS

AMPS	Advanced Mobile Phone Service
ANSI	American National Standard Institute
ARPA	Advanced Research Project Agency
ARPU	Average Revenue Per Unit. One indicator of a wireless business operating performance. ARPU measures the average monthly revenue generated for each customer unit.
ART	Autorite de Regulation des Telecommunications
ATM	Asynchronous Transfer Mode
CDMA 2000	Code Division Multiple Access 2000, is a radio transmission technology for the evolution of narrowband CDMAOne to third generation adding up multiple carriers.
CDMA	Code Division Multiple Access, is a method of spreading spectrum transmission for digital wireless personal communications networks that allows a large number of users simultaneously to access a single radio frequency band without interference.
CEPT	Conference on European Postal and Telecommunication
c-HTML	Compact Hypertext Markup Language. The language in which NTT DoCoMo developed its i-mode services.
CLEC	Competitive Local Exchange Carrier
COLT	City of London Telecommunications
DGPT	Direction Generale des Postes et des Telecommunications
DGT	Direction Generale des Telecommunications
DWDM	Dense Wavelength Division Multiplexing
EBN	European Backbone Network
EDGE	Enhanced Data for Global Evolution, is a faster derivative of GSM. It enables multimedia and broadband functions to be performed on mobile phones.
ETSI	European Telecommunications Standardization Institute
FT	France Telecom
FTSE	Financial Times Stock Exchange
GPRS	General Package Radio Service. An extension for adding faster data transmission speed to GSM networks. It is a package-based technology.
GSM	Global Systems for Mobile Communication. The European Telecommunications Standardization Institute (ETSI) and various EU research programmes, such as RACE, played an important role in establishing this standard.

HDML	Handheld Markup Language, developed by Unwired Planet, preceding WAP.
HER	Hermes Raitel
HTML	Hypertext Markup Language, is the programming language used to design and present computer sites on the Internet in a graphical user interface fashion. HTML is the language used by programmers to design a Home Page for computers on the Internet as part of the World Wide Web project.
HTTP	Hypertext Transfer Protocol. The method for moving 'hypertext' files across the Internet. Requires an HTTP programme at one end and a server at the other.
ICCC	International Conference on Computer Communications
i-mode	NTT DoCoMo's Japanese mobile Internet service.
IP	Internet Protocol
IPTO	Information Processing Techniques Office
ISDN	Integrated Services Digital Network. Switched network providing end-to-end digital connection for simultaneous transmission of voice and/or data over multiple multiplexed communication channels and employing transmission that conforms to internationally defined standards. ISDN is considered to be the basis for a 'universal network' that can support almost any type of communications device or service.
ISP	Internet Service Provider. A company that provides access to the Internet.
KDG	Kiewit Diversified Group Inc.
KfW	Kreditanstalt fur Wiederaufbau
LAN	Local Area Network, is a local data network, one that is used to interconnect the computer equipment of a commercial user.
LDDC	Long Distance Discount Calling
MFS	Metropolitan Fiber Systems Communications
MPT	Ministry of Posts and Telecommunications
MSN	Microsoft Network
NMT	Nordic Mobile Telephony
PCG	Pacific Capital Group
PCM	pulse-code modulation
PDC	Personal Digital Cellular
PKS	Peter Kiewit Sons' Inc.
RBOC	Regional Bell Operating Company. Any one of the seven local telephone companies created in 1984 as part of the break-up of AT&T. The RBOCs are Ameritech, Bell Atlantic, Bell South, NYNEX, Pacific Telesis Group, Southwestern Bell, and US West.
RegTP	Regulatory Authority for Telecommunications and Posts (Germany)
SDH	Synchronous Digital Hierarchy
SMS	Short Message Service. Method within GSM-telephony for sending short messages from and to mobile phones.
SPC	Stored Program Control
TCP	Transmission Control Protocol
TDMA	Time Division Multiple Access, is a method of digital transmission for wireless telecommunications systems that allows a large number of users simultaneously to access a single radio frequency band without interference.
TMA	Telecommunications Managers' Association
UMTS	Universal Mobile Communications System. The third generation mobile phone system.

URL   Uniformed Resource Locator. The standard way to give the address of any resource that is on the Internet and is part of the World Wide Web.

VOD   Video-on-demand

WAN   Wide Area Network, is a data network used to interconnect a company's remote site or widely-dispersed computer equipment.

WAP   Wireless Application Protocol

WCDMA   Wideband Code Division Multiple Access. One form of multiple access in the wireless communication field. The basic technology does not differ from CDMA, but WCDMA uses broader frequency bandwidth waves.

W-LAN   Wireless Local Area Network, commonly used in offices in the US, is seen as an infrastructure that has the potential to compete with 3G networks by offering local access to users of mobile devices.

Introduction

The opening years of the twenty first century will be remembered for two events: the collapse of the Twin Towers in New York and the biggest financial collapse since the end of the Second World War. This book begins with the financial collapse.

THE TELECOMS BOOM AND BUST, 1996–2002

More specifically, this book begins with the Telecoms Boom and Bust, 1996–2002. Between 1996 and 2001 a total of $1,805 billion was invested in the Telecoms Industry through financial markets. At the height of the telecoms boom in March 2000 the total stock market value of all telecoms operators and equipment suppliers was $6,300 billion. By September 2001 (just before the collapse of the Twin Towers) the value had fallen to $3,800 billion, a fall of $2,500 billion. By contrast, the loss in stock market value in all the Asian markets combined in the Asian Crisis of 1997/8 amounted to $813 billion.[1]

Why did the telecoms boom occur in the first place? What were the major forces that drove it? Why did the boom turn into a bust? Who, if anyone, was to blame? What were the consequences for the Telecoms Industry and its companies (that included household names such as AT&T, BT, Cisco, COLT, Deutsche Telekom, Ericsson, France Telecom, Fujitsu, Lucent, NEC, Nortel, Nokia, NTT, Qwest, Vodafone, and WorldCom)? These are the kinds of questions that are analysed in Chapter 1.

THE TELECOMS INDUSTRY, THE INTERNET, AND THE INFOCOMMUNICATIONS INDUSTRY

Important though it is, the analysis of the Telecoms Boom and Bust in Chapter 1 only captures the tip of the iceberg that in this book is referred to as the

[1] *Financial Times*, 5 September 2001.

Infocommunications Industry. It is the Infocommunications Industry that brings together telecoms services and equipment, computers, software, semiconductors, the Internet, e-commerce, and some of the media. This is the industry that has been the most important engine of growth in the world economy since the Second World War. It is also the Infocommunications Industry that has created the opportunity for the emergence of new postwar firms (such as Cisco, Intel, Microsoft, Nokia and Vodafone), new industrial districts (like Silicon Valley and its lesser counterparts), and new industrializing countries (including Japan, Korea, and Taiwan).

How has the Telecoms Industry contributed to the emergence of the Infocommunications Industry? How, precisely, does the Telecoms Industry 'fit into' the Infocommunications Industry? What role is played by the Internet in this industry? Most complex of all, what are the forces that drive the evolution of the Infocommunications Industry? As these questions make clear, a more fundamental analysis is needed, one that goes beyond the explanation of the Telecoms Boom and Bust, in order to understand the dynamics of the Infocommunications Industry. In Chapter 2 these questions are answered.

THE INCUMBENTS

Until the 1980s, state-controlled monopolies—such as the Big Five: AT&T, BT, Deutsche Telekom, France Telecom, and NTT—assisted by specialist telecoms equipment companies, provided a nation's telecoms services. Thereafter began an era of liberalization—starting in the mid-1980s in Japan, the UK, and the US—when the state-controlled monopolies were forced to confront new competitors and all telecoms markets were opened to a greater amount of competition.[2] At the same time the Telecoms Industry was rocked by creative–destructive change as new technologies permeated the industry. These new technologies included packet-switched data networks, optical switching and transmissions systems, the Internet, and mobile communications.

In many other industries such radical discontinuities in both the knowledge base and the competitive environment have shaken the structure of the industry, opening the doors to new entrants who, riding on the new technologies, have challenged the incumbents and in some cases replaced them. The computer industry after the advent of the microprocessor and the PC is an example.[3]

How has the Telecoms Industry in the Internet Age adapted to the radical changes transforming its knowledge base? Have the incumbent network

---

[2] The European Union officially opened all its telecoms markets to competition from 1 January 1998.   [3] See, for example, Grove (1996) and Mowery and Nelson (1999).

operators been super-adapters, successfully mastering the new technologies, entering the new business areas, and transforming the old, or have they been dinosaurs in an age of climate change? More specifically, how have they coped with the radical changes brought about by the emergence of mobile communications, the Internet, and data communications?

In Chapter 3 a comparative analysis is provided of the three incumbent network operators who were the first to be forced, in the mid-1980s, into the era of liberalization and competition, namely AT&T, BT, and NTT. How have they dealt with the forces driving the industry, analysed in Chapters 1 and 2? Which of the three have performed best and in which areas? What changes have they had to make in order to survive and prosper in the evolving Infocommunications Industry? In Chapters 4 and 5 the same kinds of questions are asked about the two largest European incumbent network operators, Deutsche Telekom and France Telecom.

THE NEW ENTRANTS

The era of telecoms liberalization that began in the mid-1980s unleashed new forces by creating new opportunities for entrepreneurs and companies to make money in an industry that hitherto had been the privileged preserve of the state-controlled monopoly operators. Some of the richest people in the US—such as Philip Anschutz who established Qwest and George Soros who was an investor in GTS—were attracted by the perceived bonanza. The new opportunities in telecoms also provided openings for completely new entrepreneurs like Bernard Ebbers, the former football coach and motel operator, who with colleagues established the company that by the late 1990s, according to some financial analysts, posed the biggest threat to the incumbent AT&T, namely WorldCom. In Europe, the US, and Japan established companies—such as utility companies, financial institutions, and industrial companies—invested in new entrant operators in order to share in the expected profits.

In many cases, the entrepreneurs and companies that established new entrant telecoms operators had little prior knowledge of the telecoms industry and its technologies. This raises an important question: how could they enter an industry of which they had so little knowledge? What does the experience of the entrepreneurs and their new entrant companies tell us about the dynamics of the Telecoms Industry in the Internet Age? These questions are answered in Chapter 6 for five of the most important new entrants in the US—Global Crossing, GTS, Level 3, Qwest, and WorldCom.

For reasons discussed in Chapter 1, many leading financial analysts in the mid-1990s put their money on the new entrant operators rather than the incumbents. Their faith in the new entrants contributed to the Telecoms Boom. But how well have the new entrants tended to do and have they lived up to expectations? This question is dealt with in a general way in Chapter 1. However, in Chapter 7 it is dealt with far more specifically through a detailed analysis of the fortunes of what many financial analysts argue is the most successful of all the new entrants in Europe, namely COLT. COLT was established in London in 1992 by Fidelity, the largest US mutual fund. In 1999 COLT was the best performing company on the London Stock Exchange, and its President and CEO the best paid executive in the UK. Why was COLT able to enter the European telecoms industry so speedily and successfully? How was the company affected by the Telecoms Bust? What are its future prospects? These key questions are examined in Chapter 7.

THE SHIFTING KNOWLEDGE-BASE OF THE TELECOMS INDUSTRY

In the old Telecoms Industry, before the era of liberalization and competition, the innovative engine of the industry was firmly located in the famous central research and development laboratories of the leading incumbent operators—such as AT&T's Bell Laboratories, BT's Martlesham Laboratories, France Telecom's CNET, and NTT's Electrical Communications Laboratories. These laboratories—particularly Bell Labs that boasted more Nobel Prize winning researchers than any other industrial laboratory—produced the technologies that would drive, not only the Telecoms Industry itself, but also related industries such as computers, software, and consumer electronics.

However, powerful though these central R&D labs were as engines of innovation, they suffered from one significant drawback: they were closed systems. Although the incumbent operators were able to attract many brilliant researchers to their laboratories, entry into the telecoms innovation system was highly limited. In effect, entry was restricted to the employees of the incumbent operators and their closed group consisting of several telecoms equipment suppliers.

From around the 1970s, however, this knowledge-creating system in the Telecoms Industry was to change radically. Two forces drove the change. The first was the specialized knowledge gradually accumulated by the specialist telecoms equipment companies—old established companies such as Lucent, Nortel, Ericsson, NEC, and Fujitsu and new companies like Cisco and Nokia. Increasingly, these companies researched, developed, and manufactured the 'network elements' that were purchased by the network operators who turned them into networks which they used to provide telecoms services. Unlike the incumbent network operators in the monopoly era, the specialist telecoms equipment suppliers faced competition, limited in the large industrialized countries by the long-term obligational relationships that bound them to the national incumbent, but much fiercer in other countries including the developing world. To a significant extent they competed by innovating.

The second driver of change in the telecoms knowledge-creating system was the Internet that introduced an entirely new system. In contrast to the old system, the new system created by the Internet was open, had low entry barriers, admitted many innovators, had a common knowledge base, provided high-powered incentives (both material and non-material), and produced rapid, concurrent, innovation. These two drivers of change are analysed in detail in Chapter 2.

The changing knowledge-base of the Telecoms Industry in the Internet Age is evident in the changing location of R&D in the industry which is examined in Chapter 2 and in the sections on R&D in Chapters 3, 4, and 5 on the incumbents. The essence of the change can be summarized in one table, Table 8.2, Chapter 8. This shows two remarkable facts. First, the Big Five incumbent network operators are on average about as R&D-intensive as the beverages industry, and less R&D-intensive than industries not normally considered as 'high tech', such as personal care, media and photography, and automobiles. Secondly, Table 8.2 shows that the specialist telecoms equipment companies are four to five times more R&D-intensive than the Big Five incumbents. Indeed, AT&T allocated only 0.9 per cent of its sales to R&D. This compared with 15.4 per cent for Ericsson, the most R&D-intensive of the specialist equipment suppliers included in the table. In short, Table 8.2 shows that very little R&D needs to be done by network operators wanting to enter and survive in the Telecoms Industry.

This shift in the knowledge base of the Telecoms Industry has had extremely important implications for the dynamics of the entire industry. More specifically, the shift has facilitated (and reflects) a process of vertical specialization, that is, a division of labour between R&D-intensive specialist equipment suppliers on the one hand and non-R&D-intensive network operators on the other. In turn, this has drastically lowered entry barriers into the network operator layer of the Telecoms Industry. The important implications that this has had for the Telecoms Boom and Bust are analysed in Chapter 1 while the crucial general implications for the evolution and dynamics of the industry are examined throughout the book.

THE MOBILE INTERNET

One of the most important recent events in the Telecoms Industry in the Internet Age has been the convergence of mobile communications and the Internet in the form of mobile Internet access and services. The mobile Internet also presents a fascinating puzzle. At about the same time, in the late 1990s, attempts were made both in Japan and in Europe to reformat web pages so that they could be accessed by mobile phones.

The Japanese attempts were first made by NTT DoCoMo, the mobile subsidiary of the incumbent operator, NTT. By 2002 DoCoMo's i-mode mobile internet service had over 30 million customers in Japan. The European attempt was spearheaded by Ericsson, Nokia, and Motorola and was called wireless application protocol (WAP). However, while i-mode was a resounding success, WAP was a dismal failure (though more is hoped from later generations of this technology). How is this significant difference in outcome to be explained? This question, which has important implications both for the future of the Internet and mobile communications, is answered in Chapter 9.

CONCLUSIONS

What are the conclusions that can be drawn from the present studies of Telecoms in the Internet Age? This final question is dealt with in the concluding section at the end of this book.

TELECOMVISIONS.COM

The author, with colleagues,[4] has developed an interactive web site dedicated to analysing what will happen to the Telecoms Industry in the Internet Age and its companies over the next five years. This web site contains a good deal of analysis and information which complements that in the present book. The site may be located at: http://www.TelecomVisions.com.

[4] Sadahiko Kano, Tim Bolt, Lynne Dyer, and Ian Duff.

1

Explaining the Telecoms Boom and Bust, 1996–2002

This introductory chapter sets the scene for the entire book by situating the Telecommunications (Telecoms) Industry in the Internet Age within the context of the remarkable financial boom and bust that occurred between 1996 and 2002. This boom and bust created enormous waves that swept the industry along until March 2000, allowing the leading telecoms companies to ride the waves to new heights but thereafter dumping them into new depths.

Why did the Telecoms Boom occur? What were the forces that drove it? Why did it end, turning into a gigantic bust, the consequences of which are still shaking the whole industry? And who, if anyone, is to blame? These are some of the questions that are analysed in this chapter.

Equally importantly, what are the consequences of the boom and bust? Specifically, how has the boom and bust affected the incumbent telecoms operators (such as AT&T, BT, Deutsche Telekom, France Telecom, and NTT), new entrants (like WorldCom and Colt), telecoms equipment specialists (including Nortel, Lucent, Cisco, Ericsson, and Nokia), and mobile specialists (notably Vodafone)? This question is analysed in a general way in this chapter and is then examined in greater detail in subsequent chapters that delve far more deeply into the impact on some of these companies as well as their responses to the boom and bust.

INTRODUCTION

The telecoms boom and bust of 1996–2002 has been the most remarkable industrial collapse in postwar history. The figures supporting this claim make impressive reading. Between 1996 and 2001 syndicated bank loans to telecoms companies totalled $890 billion, private equity and stock markets contributed a further $500 billion, while bond markets provided $415 billion of debt, making a total of $1,805 billion. In March 2000, at the height of the telecoms boom, the total stock market value of all telecoms operators and equipment suppliers was $6,300 billion; by September 2001 the value had fallen to $3,800 billion, a fall of $2,500 billion. In contrast, the loss in stock market value in all the Asian markets combined in the Asian crisis of 1997/8 amounted to $813 billion.[1]

As always, some of those picking through the ashes after a collapse were quick, with the benefit of hindsight, to assign blame. A particularly noteworthy example was Jack Grubman, the key telecoms analyst at Salomon Brothers. Grubman became one of the main prophets supporting the new telecoms operators who entered the telecoms industry to challenge the incumbents, therefore making an important contribution to what became known as the Telecoms Bubble.[2] According to Grubman,

Everyone was culpable [for the telecoms bubble and its bursting]: the debt markets, the equity markets, the issuers, the companies. Anyone who tries to point a finger at a single group is not being fair.[3]

But this blame raises more questions than it answers. If everyone was 'culpable', this implies that they could have, and should have, known better. But if they could have, and should have, known better, why did they not know better? Where were those that did know better and either publicly counselled an alternative course of action, or privately profited, using their superior knowledge to make superior investment decisions? And, if it is hard to find evidence of significant numbers of dissenting voices, does this not contradict the earlier statement that everyone could have, and should have, known better? As these questions clearly imply, providing an explanation of the Telecoms Bubble is more difficult than might at first sight appear.

The Argument in this Chapter

The first step will be to re-construct the Consensual Vision (or cognitive framework) that shaped most of the thinking and decision-making that drove the Telecoms Boom from around 1996–2000. (A 'vision' or cognitive framework consists of an interrelated set of beliefs, embodied in assumptions and

---

[1] *Financial Times*, 5 September 2001.

[2] Some relevant quotations from Grubman are provided later in this chapter in the subsections titled 'Consequences for the New Entrant Telecoms Operators' and 'Consequences for the Mobile Companies'.

[3] *Financial Times*, 7 September 2001. Later, *Business Week* was to note that 'At the height of the bull market, Salomon Smith Barney telecom analyst Jack Grubman had buy recommendations on practically all the companies he covered. During that time, Salomon was loading up on investment banking fees from telecom companies, racking up almost $1 billion since 1997—more than any other Wall Street firm. Now, nine of the companies Grubman cheered during the telecom craze are trading for less than $1 a share. At least four are in bankruptcy.' (25 February 2002, p. 47.)

expectations, which serve the purpose of making the world seem intelligible and therefore orienting decision-making.)[4]

The second step will be, with the aid of hindsight, to examine some of the flaws in the Consensual Vision. Having done this we shall, third, return to the question of whether key decision-makers could have, and should have, known better and, if so, why they did not.

The Consensual Vision in Telecoms[5]

Stripped to its essentials, there were four planks in the Consensual Vision in telecoms.

Explosive Demand for Bandwidth. The central plank was the expectation of 'explosive demand for bandwidth'. What were the causal factors that shaped this expectation?

The answer to this question can be encapsulated in one word: Internet. In Chapter 2 of this book a detailed analysis is provided of the complex evolution of the Internet and its impact on the Telecoms Industry based on its triad of technologies—packet-switching, Internet Protocol (IP), and the World Wide Web. In Chapter 2 it is shown that it was in 1995 that Bill Gates finally realized that it was the Internet, rather than Microsoft, that would shape the world of computing. The rapid diffusion of the Internet had already begun.

In the same year, Gates was invited by Warren Buffett, the investor 'sage of Omaha', to a private meeting for investors in Dublin. At this meeting Gates spoke about his newfound belief in the potential of the Internet. Attending the meeting was one Walter Scott Jr, an engineer who had for decades been Chairman of Peter Kiewit Sons', a Nebraskan conglomerate that had been primarily involved in construction. Hearing Gates, Scott came to the conclusion that the Internet was nothing more than a network of IP networks. This insight led him, eventually, to create Level 3, one of a new generation of aggressive new

---

[4] For academic writing on visions and cognitive frameworks, see Fransman (1999), March (1994), Loasby (2000, 2001a,b), and Witt (2000, 2001).

[5] There were, of course, a number of precipitating factors that created the conditions that allowed the Telecoms Boom to emerge. These factors included the changes in legislation in the US that gave people more control over the investment of their pension funds and allowed them to hold a greater proportion of equity, the increasing proportion of household savings held in the form of equity after the bull market made its appearance in the early 1980s, and the concomitant rise in the significance of mutual funds in the US and their equivalent in other countries as an increasingly important category of institutional investor. However, it is still necessary to explain why it was the Telecoms Industry that was perceived to provide one of the most promising areas for investment, thus initiating the Telecoms Boom. It is this explanation that is the main focus of the present chapter, rather than the precipitating factors. For a description of the precipitating factors, and an account of the rise and fall of the so-called dot com companies, see John Cassidy, *dot.con* (Allen Lane 2002).

entrant telecoms operators that built their networks on IP technology. Others, at the same time, quickly came to a similar conclusion.[6] The link between the Internet and demand for telecoms bandwidth had been forged.

New Telecoms Operators will Out-compete the Incumbents. But demand is one thing, opportunities for relatively profitable investment another. More specifically, who would be best placed to meet this demand: the incumbents (such as AT&T, BT, Deutsche Telekom, France Telecom, and NTT) or the new entrant telecoms operators? Would demand increase fast enough for both these groups of companies to grow rapidly, or would a battle ensue with only one group emerging as victor? It was in answering this question that Grubman and his colleagues at Salomon Brothers, together with others of similar persuasion, made their contribution to shaping the second plank of the Consensual Vision: that the fittest new entrant telecoms operators would enjoy a competitive edge over the incumbents. This contribution came in the form of a number of studies of new entrant telecoms operators that included companies such as WorldCom, Qwest, and Level 3 in the US, and COLT and Energis in Europe.[7]

In turn, this second plank was based on four other crucial assumptions. The first assumption was that the incumbents, though enjoying advantages of economies of scale and scope, existing market domination, financial muscle, and brand recognition, were severely constrained technology-wise by their legacy networks.[8] Second, the technology constraint on incumbents created a window of opportunity for the new entrants who could rapidly deploy the latest technologies that were available on technology markets. In this way they could seize a technological competitive edge in their battle with the incumbents. (In the studies by Salomon Brothers, for example, much was made of the deployment by the new entrants of new technologies such as Sonet/SDH and self-healing ring architectures that facilitated higher bandwidth and quality of service.)

Third, it was argued, the new entrants also possessed organizational advantages over their incumbent rivals. While the new entrants were small, focused, flexible, fast, and flat, their opponents were large, complex, hierarchical, bureaucratic, and prone to inertia. Fourth, while the incumbents were impeded by regulation and universal service obligations, the new entrants could cream-skim. By concentrating investments in only the high-usage parts of the major

---

[6] For more details on Level 3, including this information on Scott, and some of the other new entrant telecoms operators, see the articles prepared under the present author's direction in the Articles Zone of http://www.TelecomVisions.com.

[7] A general discussion on the approach used in these kinds of company studies and its shortcomings is to be found in the subsection 'Inventing the Value of Telecoms Company Shares' in Chapter 2. Some of the Salomon studies are referred to in the bibliography under J. B. Grubman and A. Harrington.

[8] Legacy networks refer to the incumbents' existing networks embodying older generations of technology.

cities (e.g. financial districts), with relatively small investments and relatively little geographical coverage the new entrants could address the major proportion of the rapidly-growing telecoms services markets, such as that for corporate data, which was quickly outstripping the market for voice services.

These four assumptions led to the crucial conclusion that the new entrants would outperform the incumbents in meeting the explosive demand for bandwidth. In turn, this conclusion led to the third plank in the Consensual Vision.

Financial Markets will Support the Fittest New Entrants. The third plank of the Consensual Vision was that financial markets would give significant support to the fittest new entrant telecoms operators relative to the incumbents. To the extent that this expectation was realized in reality, the market value of the assets of these new entrants would rise, resulting in relatively attractive returns to investors.

In turn, this led to the establishment of a positive feedback loop. Financial markets (debt, equity, and bank-lending), accepting the planks in the Consensual Vision, supported the fittest new entrants such as WorldCom, Qwest, Level 3, Global Crossing, COLT, and Energis. As a result of this support, the market value of these companies rose, particularly after their IPOs. This led to a self-fulfilling prophecy, further fuelling the cycle of self-fulfilling expectations and justifying the 'buy' recommendations of analysts such as Grubman.

Technological Change will Further Reinforce the Chain of Expectations. The final plank in the Consensual Vision was that technological change would further reinforce the chain of expectations on which the vision was based, thus adding an endogenous element to the causation. Technological change, stimulated by demand from the rapidly growing new entrant telecoms operators, would further contribute to their competitive success against the legacy-constrained incumbents. In turn, this would increase the relative attractiveness of these new entrants. As a consequence, their relative share price would rise even further, reinforcing the original expectations. In this way the power of technological change provided further fuel for the positive feedback loops.[9]

Conclusion. These processes, shaping the formation of the Consensual Vision, also had the effect of harmonizing the interests of private investors, financial institutions, new entrant telecoms operators, equipment suppliers, and even regulators, thrilled at seeing the possibility of significant entry and competition. The harmony of interests also provided the basis for the convergence of expectations around the Consensual Vision.

---

[9] In his recent book, *Irrational Exuberance*, Robert Shiller (2000) emphasizes the role played by feedback loops in causing speculative financial bubbles that preceded the great financial collapse of 2001, see pp. 60–8.

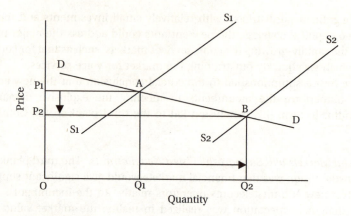

Fig. 1.1. The effect of a fall in the price of bandwidth.

Destruction in the Creation?

But were the shapers of the Consensual Vision not troubled by the possibility that new technology might be a double-edged sword, that is, also acting on the supply side of telecoms services markets by increasing capacity and lowering both prices and profits? Could not technology, in true Schumpeterian fashion, be a destructive as well as a creative force?

This possibility was not lost on some of the shapers of the Consensual Vision. However, they were soon provided with an additional analytical weapon in their armoury, supporting the vision, the elasticity of demand.[10] This argument is presented in a simple diagram in Fig. 1.1.

As Fig. 1.1 clearly shows, even if the most pessimistic assumptions are made (i.e. demand for bandwidth remains constant while supply increases significantly from S_1S_1 to S_2S_2), total revenue (i.e. market size) will increase, as long as the demand curve is elastic. Since the purchasers of bandwidth increase the quantity of bandwidth bought (from Q_1 to Q_2) as its price falls (from P_1 to P_2), the new total revenue (OP_2BQ_2) significantly exceeds the old total revenue (OP_1AQ_1).

However, the key plank in the Consensual Vision argued that, far from remaining constant as supply increased, the demand curve would shift outwards rapidly. This would occur as new Internet-related bandwidth-intensive applications—such as video-enabled web sites and video-based instant messaging—emerged. The outcome, it was strongly believed, would be a rapid increase in the size of the market available to telecoms operators and service providers. In turn, this would facilitate rapid growth in the revenue and profitability of the

---

[10] In numerous conferences and interviews the present author heard the elasticity of demand argument in defence of the Consensual Vision. (If demand for bandwidth is elastic then a proportional fall in the price of bandwidth will result in a greater proportional rise in demand. This implies a rise in total revenue.)

fittest, new entrant telecoms operators, particularly given their competitiveness *vis-à-vis* the incumbents. In this way assumptions about the elasticity of demand further reinforced the optimistic view regarding the potential of the fittest new entrants put forward by the Consensual Vision.

Evidence supporting this interpretation is provided by data on the relative movement in the share prices of the incumbents and two of their new entrant competitors—AT&T, and Qwest and WorldCom in the US, and BT, and COLT and Energis in the UK. This data is shown in Fig. 1.2.

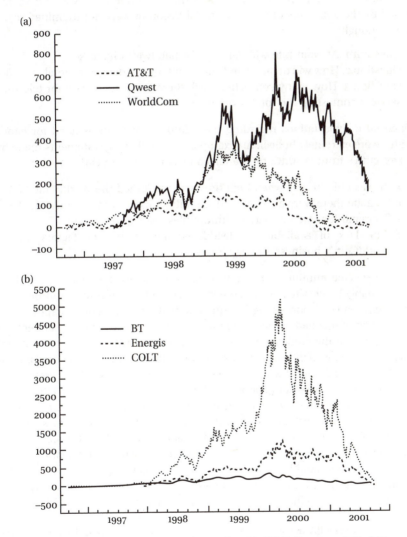

Fig. 1.2. Incumbents and new entrants: share price. (a) AT&T, Qwest, and WorldCom: share price; (b) BT, COLT, and Energis: share price.

As can be seen from Fig. 1.2, from 1997 the share price of the main new entrants began to significantly exceed that of the incumbents, thus confirming the argument of the Consensual Vision. This continued until the latter part of 2001 when the gap considerably narrowed or was even eliminated.[11]

Flaws in the Consensual Vision

With hindsight, however, it is clear that three key questions were either not posed by the adherents of the Consensual Vision or were not examined rigorously enough:

Question 1. At what rate will demand for bandwidth increase?
Question 2. How will demand and supply interact to determine total revenue?
Question 3. How many competitors will there be in the market for telecoms services contesting the total revenue available?

Question 1. Expectations regarding the rate of increase in demand for bandwidth were extremely bullish. An example is the following statements made by an executive from WorldCom at a telecoms conference in early 2000:

- 'It was only in 1994 when Web browsers unleashed the WorldWide Web.'
- 'In 2000 there were an estimated 163 million users of the WorldWide Web.'
- 'Demand for Internet bandwidth doubles every 3 to 4 months.'
- 'The lack of available bandwidth is one of the key factors driving WorldCom's business strategy today.'

The expected imminent arrival of bandwidth-intensive applications such as video-enabled web sites, video-conferencing, and video-on-demand over the Internet seemed to add strength to the optimistic forecasts of demand.

However, if one had asked at the time: 'What evidence is there that demand will increase at the predicted rates?', the only true answer could have been that there was little evidence to support the predictions. Although the demand for bandwidth at the time was increasing rapidly, there was no necessary reason for it to continue doing so at the same rate. For one thing, the saturation effect associated with the normal sigmoid diffusion curve had not been taken into account. For another, the expected speed with which innovative new services such as video-enabled web sites would be widely adopted was based more on wish than on reality. In short, the available information left a significant degree of interpretive ambiguity[12] regarding what would happen to the demand for bandwidth in the future. Into the 'holes' left in the available information set by ambiguity seeped bullish expectations shaped by the Consensual Vision. The

---

[11] The reasons for this in the case of AT&T and WorldCom are analysed in detail in Chapter 3.
[12] For a more detailed discussion of the concept of 'interpretive ambiguity', see Fransman, *Visions of Innovation* (1999).

'knowledge'—'telecoms was a good bet'—that the decision-makers, whose decisions drove the Infocommunications Industry, thought they had was derived more from the set of consensual beliefs than from an unambiguous set of data.

We now know with hindsight that although the demand for bandwidth, driven by the demand for data services of various kinds, did increase rapidly, it did not increase as rapidly as had been forecast. One reason has been the slower-than-expected rollout of broadband connectivity to the home and small businesses. This has partly been the result of a lack of competition in the local loop (the 'last mile' connecting homes and businesses) due to incumbent intransigence, the lack of effective regulatory tools, and the lack of sufficiently enticing broadband services, content, and applications, themselves the result of limited broadband take-up by the mass market.

Question 2. In retrospect it is also surprising how frequently supporters of the Consensual Vision referred to demand predictions in support of their strategies but how seldom they integrated predictions about supply (the statements from WorldCom quoted earlier being a case in point). In this connection one is reminded of the observation made by the nineteenth century Cambridge (UK) economist, Alfred Marshall, who noted that demand and supply are like the two blades of a pair of scissors; just as the two blades are necessary to cut paper, so demand and supply necessarily jointly determine price; and, of course, price is a determinant of revenue and, therefore, profit. The supporters of the Consensual Vision, however, often tried to cut paper with only one blade.

Had they looked more closely, they would have found ample evidence of a rapid increase in the supply of bandwidth.[13] Ironically, for example, at the same conference at which the WorldCom executive just referred to made his presentation, another presentation (inevitably, by someone on the 'technical side') contained the following observations:

1. 'The carrying-capacity of optical fibre is doubling every twelve months'. (The famous Moore's Law in semiconductors predicted a 'doubling of transistor density on a manufactured die [chip] every year'—Gordon Moore, Intel, 1965.)
2. 'The carrying-capacity of wireless channels is doubling every nine months'.
3. 'Lucent Technologies' Bell Labs has pushed 1.6 trillion bits, or terabits, of information through a single fibre by using the dense wavelength division multiplexing technique [DWDM].[14] This is enough for 25 million conversations or 200,000 video signals simultaneously and one cable may contain a dozen such fibres!'

[13] More accurately, they did 'look', but failed to integrate what they saw into a satisfactory analytical framework. This is a good example of cognitive framework determining the set of information used in tackling a problem, rather than the other way round.

[14] DWDM is an optical technology that uses different coloured light waves as communication channels in a single optical fibre, thereby increasing significantly the carrying-capacity of that fibre.

Surprisingly, the 'two blades of the scissors' were separated at the telecoms conference at which these two sets of statements were made and no attempt was made to bring them together. To some extent the possible negative effects of an increase in supply are anticipated in the elasticity of demand argument outlined in Fig. 1.1. However, there is a limit to the argument if supply increases, and price falls, sufficiently.

Naturally, there were some voices in the wilderness sounding an early alarm warning. One example was the argument about the coming 'bandwidth capacity glut' put forward by a small consultancy in Boston, Massachusetts, and given publicity through telecoms articles in the *Financial Times*.[15] But perhaps precisely because such warnings challenged the Consensual Vision and raised uncomfortable questions that might have been damaging to confront, they tended to be ignored. In the event, however, by September 2001 it was estimated that 'only 1 or 2 percent of the fibre optic cable buried under Europe and North America has even been turned on or "lit"....A similar overcapacity exists in undersea links, where each new Atlantic cable adds as much bandwidth as all the previous infrastructure put together'.[16,17]

Question 3. However, the Achilles heel of the Consensual Vision lay in its failure to understand the changing structure of the Telecoms Industry, as it metamorphosed into the new Infocommunications Industry, and the implications of these changes. One of the main contributions of the present book is its analysis of the evolution of industrial structure as this metamorphosis took place.

An appreciation of the changes that had taken place in the structure of the Telecoms Industry would have revealed that the opportunity for above-average profits in the telecoms network layer of the industry was not nearly as great as portrayed in the Consensual Vision. The reason was that in the new Infocommunications Industry the barriers to entry into the network layer had been dramatically lowered in many key market segments.[18] As a result, a significant

---

[15]   See Tables in Chapter 3 (comparing AT&T, BT, and NTT).

[16]   *Financial Times*, 4 September 2001.

[17]   Writing in April 1998, Jack Grubman produced a glowing report on WorldCom, calling the company in the title of the publication 'the only legitimate telecom large-cap growth stock'. In this publication Grubman stated: 'Since we continue to believe that there will be a chronic shortage of bandwidth, ownership of end-to-end bandwidth [such as WorldCom possessed] is key to having long-term, profitable growth'. However, Grubman did not provide investors with the reasoning behind this key belief and failed to analyse the forces on the supply side that might have undermined it. Jack B. Grubman and Sheri McMahon, 'WorldCom, Inc. Combination with MCI Creates the Only Legitimate Telecom Large-Cap Growth Stock', *Salomon Smith Barney Company Report*, 9 April 1998, p. 7.

[18]   It must be emphasized, however, that barriers to entry were not dramatically lowered in all market segments. For example, in the local access market (the market providing connectivity to homes and businesses) the costs of entry by new competitors remained high. This was the result of high investment costs, opposition from the monopoly incumbent, and the difficulties that regulators confronted in opening these markets. Similarly, in mobile communications, barriers to entry also remained high as a result of the relatively small number of mobile licences that were given by most governments. In strong contrast, however, barriers to entry into

number of aggressive network-based new entrants were able to rapidly enter the industry. In turn, this meant that in those market segments where the new entrants entered, the supply of network capacity and network services increased rapidly. This led to falling prices. It also meant that the total revenue generated in these segments would have to be shared amongst a larger number of competitors.

In effect, this negated to some extent the optimistic conclusions that the supporters of the Consensual Vision derived from the elasticity of demand argument summarized in Fig. 1.1. For even if total revenue were to increase consequent upon a fall in price, the larger number of competitors could imply that the growth of revenue, or even absolute revenue, per firm or for an individual firm, could fall. The same outcome could result even if the demand for bandwidth increased significantly. Indeed, as is shown in detail in Chapters 3–6, this is precisely the problem that AT&T, BT, Deutsche Telekom, France Telecom, and NTT came to face. Chapter 3 also shows that this difficulty plagued not only the incumbents but also the new entrant, which until 2000/2001 had been by far the most successful of all the new entrants in the US, WorldCom.

But these consequences all hinge on the fall in barriers to entry which facilitated a sharp increase in the intensity of competition in the related market segments.[19] This raises the key question regarding why barriers to entry fell. This important question is analysed in the next subsection.

The Vertically-specialized Infocommunications Industry. In Chapter 2, a detailed analysis is provided of the evolution of the structure of the Telecoms Industry as it was transformed by the advent of the Internet into the new Infocommunications Industry. This analysis constitutes one of the main contributions of the present book. In this subsection a summary will be provided of the processes that led to the sharp fall in the barriers to entry into the network layer of the Infocommunications Industry.

Knowledge and the Infocommunications industry. From the account given thus far of the process of boom and bust in the global Telecoms Industry it is clear that a major problem was presented by limited knowledge of how the industry worked, how it was changing, and what would happen to key parameters such as the demand and supply of bandwidth and new services, and the extent of new entry. We have seen how knowledge of these kinds of issues was embodied in the cognitive framework of the Consensual Vision but we have also examined some of the limitations of this 'knowledge'. However, knowledge also enters into the story of the Telecoms boom and bust in a fundamentally different way, as a productive input that facilitates output—this is also important in explaining the boom and bust.

---

markets for long-distance and international voice and data were significantly lowered with effects that are discussed in this chapter and throughout this book.

[19] See previous footnote.

Table 1.1. The layer model of the Infocommunications Industry

Layer	Activity	Example companies
VI	*Customers*	—
V	*Applications Layer, including contents packaging* (e.g. Web design, on-line information services, broadcasting services, etc.)	Bloombergs, Reuters, AOL-Time Warner, MSN, Newscorp, etc.
IV	*Navigation & Middleware Layer* (e.g. browsers, portals, search engines, directory assistance, security, electronic payment, etc.)	Yahoo, Netscape, etc.
III	*Connectivity Layer* (e.g. Internet access, Web hosting)	IAPs and ISPs
IP Interface		
II	*Network Layer* (e.g. optical fibre network, DSL local network, radio access network, Ethernet, frame relay, ISDN, ATM, etc.)	AT&T, BT, NTT, WorldCom, Qwest, Colt, Energis, etc.
I	*Equipment & Software Layer* (e.g. switches, transmission equipment, routers, servers, CPE, billing software, etc.)	Nortel, Lucent, Cisco, Nokia, etc.

In order to elaborate, it is worth reproducing the layer model of the Info-communications Industry which is analysed in detail in Chapter 2. This is done in Table 1.1.[20]

The layer model. Table 1.1 provides a model of the Infocommunications Industry. Drawing on the layer models used by telecoms engineers, it shows how the Infocommunications Industry comprises a modularized system consisting of six hierarchically structured, and functionally interdependent, layers. Each higher layer is built on a 'platform' comprising the lower layers. At a more disaggregated level, each layer is itself divided further into modules, sub-modules, sub-sub-modules, etc. A layered, modularized architecture creates a powerful potential for innovation (provided the incentives are sufficient) precisely because it facilitates the division of knowledge. Those who are involved in developing and improving a module need only have a limited knowledge of what goes on in functionally related modules, as long as they have knowledge of the interfaces connecting the modules. The division of knowledge in this way assists in the construction of highly complex systems of which the Info-communications Industry is a prime example.[21]

---

[20] I would like to acknowledge the contribution that Sadahiko Kano made in developing this version of the Layer Model. For further discussion on the Layer Model see http://www. TelecomVisions.com.

[21] For a detailed examination of the significance of modularization and an application to the computer industry see Baldwin and Clark (2000). For an analysis of the economic and organizational implications of modules and networks, see Langlois and Robertson (1995).

Traditionally (i.e. in the days when voice was the main telecoms service and before the data and Internet revolutions), the Telecoms Industry comprised Layers 1 and 2, that is the equipment and network layers. However, with the emergence of the Internet, the Telecoms Industry was fundamentally transformed as is shown in Table 1.1. The process of evolution involved in this transformation is analysed in detail in Chapter 2.

The advent of IP (one of the Internet's triad of technologies together with packet-switching and the World Wide Web) transformed the Telecoms Industry by performing two functions in the system. First, it facilitates communication across technically diverse networks. Second, it allows the layers above the IP interface to be technically and organizationally separated from the layers below, thus allowing firms to specialize in the former layers without necessarily being involved in the latter. This has allowed the emergence of three new layers: Layer 3, the connectivity layer, Layer 4, the navigation and middleware layer, and Layer 5, the content and applications layer. In Table 1.1 a further layer has been added, Layer 6, the consumer layer, in order to emphasize that the process of consumption must also be considered as an important part of the system, necessitating an understanding of the co-evolution of consumption.[22]

The emergence of the Infocommunications Industry has brought into the system firms from other industries previously considered to be separate from the Telecoms Industry. This process of incorporation is often described as 'convergence'. In this way, the computer, semiconductor, and software industries, together with the traditional Telecoms Industry, have become part of the new Infocommunications Industry.

A major preoccupation of the present book is to examine in detail how the incumbents in the Traditional Telecoms Industry—such as AT&T, BT, Deutsche Telekom, France Telecom, and NTT—and the new entrant telecoms network operators have adapted to the Infocommunications Industry. This is the main theme of Chapters 3–8. Here, it is worth noting that although they remain firmly based in Layer 2 (the network layer) the incumbents have also entered Layers 3, 4, and 5 where they compete with non-telecoms companies.

The dynamics of the system. Having described the Infocommunications Industry in terms of the layer model, the next issue is to describe its dynamics, that is its processes of change. Obviously, this is a highly complex issue and, therefore, cannot be fully discussed here.[23] However, one key dimension of change will be mentioned here precisely because of its connection with the earlier discussion in this chapter of falling barriers to entry—the increasing intensity of competition in parts of the network layer of the industry (Layer 2), and the sharing of the total revenue generated in this layer. These issues, it will be recalled, have a key bearing on the accuracy of the Consensual Vision. Also

---

[22] For further details see Chapter 2, http://www.TelecomVisions.com and Abbate (1999).

[23] The reader is referred to Chapter 2 and http://www.TelecomVisions.com for further details.

closely related, as will be seen, is the way in which knowledge is created and used in this industry.

One of the most important organizational features of the Infocommunications Industry, important for its dynamics, is the process of vertical specialization that has evolved between Layers 1 and 2. In this process of vertical specialization, different firms have come to specialize in Layer 1 (the equipment layer that includes companies such as Lucent, Nortel, Cisco, Ericsson, Nokia, NEC, and Fujitsu) and Layer 2 (the network layer with firms like AT&T, BT, NTT, WorldCom, and Qwest).

Furthermore, the bulk of the R&D-based innovation in the industry is undertaken by the equipment specialists in Layer 1. On the other hand, the network operators in Layer 2 do relatively little R&D and, surprisingly, on average are even less R&D-intensive than the average in sectors not normally considered as 'high-tech', such as vehicles, building and materials, and leisure and hotels. (More accurately, while the incumbents such as AT&T, BT, France Telecom, and NTT still undertake a good deal of R&D in the present-day versions of their famous laboratories, their average R&D-intensity is something like a quarter of that of specialist equipment suppliers like Nortel, Lucent, etc. However, the new entrant telecoms operators like WorldCom, Qwest, and COLT have made a fundamentally different choice in deciding to outsource virtually all their R&D requirements to their chosen equipment suppliers. These issues are considered in detail in Chapters 2 and 9.)

This process of vertical specialization between Layers 1 and 2 has had several important consequences for the dynamics of the industry. First, since the specialist equipment suppliers are willing to sell their technology to whoever can pay, *all* network operators, provided they have the funding, are able to gain access to the latest technology.

Second, the technology comes in a form that requires limited knowledge on the part of the network operators. The latter's engineers have to know how to *use* the equipment—specifically, how to configure the equipment into networks and supporting front and back office systems—whilst knowing relatively little about the technology contained in the 'back boxes' being used. (In a not too dissimilar way, I can begin to use my PC or microwave oven with knowledge only of the interfaces.)

An important consequence, following from these two points, is that technological barriers to entry into Layer 2 have been relatively low. Accordingly, would-be network operators with the necessary funds and human resources have been able to easily enter Layer 2 of the industry. Furthermore, the widespread acceptance by financial markets of the Consensual Vision meant that an ample supply of capital was made available to new entrant network operators, and labour markets (often involving the head-hunting of management and skilled labour from the incumbents) made human resources available to the new entrants. All told, therefore, the barriers to entry facing potential telecoms operators were relatively low. (This point emerges clearly in Chapter 7 which analyses the role played by the entrepreneurs who engineered the entry of some of the major new entrants in the US, such as WorldCom, Qwest, Level 3, and Global Crossing. The same point emerges in Chapter 8, which examines the entry of COLT, one of the major pan-European new entrants.)

In turn, easy and rapid entry (made possible in the first place by regulatory changes) brought about an increase in the intensity of competition in parts of the Infocommunications Industry. Notable examples are long-distance and international voice and data services where, from the late 1990s, prices plummeted, causing pain to incumbents and new entrants alike. (This is shown in detail in Chapter 3, which examines the experience of AT&T, BT, NTT, and WorldCom, and in Chapters 4 and 5 where a similar process is traced for Deutsche Telekom and France Telecom.)

However, not all parts of Layer 2 have experienced the same increase in intensity of competition, followed by falling real prices. In the so-called local-loop (connecting the figurative 'last mile' between home, office, and the rest of the network) monopolistic forces have been more tenacious, despite efforts such as the 1996 Telecommunications Act in the US to dislodge them. The main reason for this relative lack of competition, however, is not technological barriers to entry, but the high investment cost of entry into the local loop compounded by efforts by the incumbents to frustrate entry. (In the US, the sagging fortunes of AT&T, in sharp contrast to the rising fortunes of the most successful Baby Bells, SBC, and Verizon, are to be explained by the former's positioning in those parts of the industry where competition has bitten due to the dynamics outlined here, and the latter's positioning in those parts still protected by other barriers to competition. See Chapter 3 for the details.)

In the case of mobile communications, the limited competition reflected in the relatively high price, for example, of calls-to-mobile, roaming, and text messaging has been more a reflection of the small number of competitors and the lack of new entrants caused by licensing restrictions. Other layers of the Infocommunications Industry have also witnessed a similar process of rapid entry by large numbers of new entrants facilitated by low technology barriers to entry. A notable case in point is Internet service providers (ISPs) which provide connectivity in Layer 3.

A further consequence of ease of access to the latest technology on the part of network operators (and others such as ISPs) is what is referred to in Chapter 2 as the 'differentiation dilemma'. This refers to the difficulty in those parts of the Infocommunications Industry affected by intense competition to use new technology as a means to differentiate their services from those of competitors in the hope of earning a scarcity rent. As in other industries, such as airlines, telecoms operators are faced with the constant dilemma of how to create the impression of differentiation in the face of an underlying technological commonality, a problem that is absent in technology-differentiation-based industries such as pharmaceuticals, microprocessors, or software (where intellectual property rights restrictions also play a key role).

Like in the airline industry, geography together with high capital costs often come to the rescue of telecoms operators despite the absence of technological differentiation. This provides a key part of the explanation for the non-globalization of Layer 2 of the industry (as documented in Chapters 3–6). In other cases, non-R&D forms of innovation have been effectively used to grapple with the differentiation dilemma. A case in point is the Orange brand, introduced by the Orange mobile operator in the UK, acquired in 2001 by France Telecom that then adopted the brand for all its mobile operations.

Vertical specialization has also been a feature of Layers 3–5. Examples are ISPs in Layer 3 (already mentioned), browsers and specialist middleware producers in Layer 4, and specialist service, content, and applications providers in Layer 5. Vertical specialization, therefore, as the present discussion makes abundantly clear, is a key determinant of the dynamics of the Infocommunications Industry. Specifically, it has facilitated rapid entry that has led to intense competition, falling prices, and a decrease in the growth of revenue and profits in the affected market segments. These trends have been one of the causes of the fall in share prices from around 2000 observable amongst the incumbents and new entrants examined in Fig. 1.2.

More Fundamentals and Some Coincidences

In the previous subsection it was concluded that the organization and use of knowledge[24] following from the process of vertical specialization in the Infocommunications Industry led to rapid entry by large numbers of new entrants, intense competition in the market segments which they entered, and falling prices, revenue, and profit growth rates. It follows from this conclusion that, even in the absence of generalized economy-wide financial and investment cycles, the boom times amongst the incumbents and new entrants in the telecoms services market (i.e. Layer 2) would have come to an end around 2000. A further 'fundamental' reinforces this conclusion.

This fundamental, a key determinant of the economics of the Infocommunications Industry, follows from the fact that many parts of this industry are characterized by high fixed and sunk costs coupled with low marginal costs. There are high fixed and sunk costs in building a telecoms network, but once it has been built the additional cost of an extra call is extremely low. Important implications follow from this characteristic.[25] The first is that in industries with this characteristic economies of scale are, by definition, important: the more a firm produces, the lower its average cost. Second, under conditions of strong competition, price will tend to be driven down towards the low marginal cost, thus eliminating supernormal profits.[26]

---

[24] This relates closely to the concepts of technology regime and learning regime developed in Chapter 2. See the related references to the literature given in Chapter 2.

[25] For a discussion, see Chapter 2 of Shapiro and Varian (1999).

[26] Shapiro and Varian (1999), pp. 21 and 24.

Shapiro and Varian (1999) suggest that 'It is this combination of low incremental costs and large scale of operation that leads to the 92 per cent gross profit margins enjoyed by Microsoft' (p. 21). However, in explaining this abnormally high margin, account also needs to be taken of the protection that Microsoft receives from its property rights over its software. Only Microsoft can produce Windows, the de facto standard for PCs. But a large number of telecoms operators can provide long-distance and international bandwidth. Although telecoms operators have a combination of low incremental costs and (in many cases) large scale of operation, as a result of low barriers to entry they also face strong competition from those competitors (in some instances incumbents from other countries) able to benefit from similar advantages. Under such competitive conditions prices tend to be pushed down to marginal cost. Accordingly, their profit margins are far lower than Microsoft's. This fundamental fact, of course, runs counter to the optimistic thrust of the Consensual Vision that shaped thinking about the prospects for the telecoms operators from around 1996 to 2000.

In addition to these internal forces, there were two other factors, external to the telecoms companies, that had a significant impact on their performance from the latter half of 2000. The first was the bursting of the so-called Dot Com Bubble, based on another Consensual Vision that also drew bullish implications from the mass take-up of the Internet. The second factor, that was related, was the downturn in the investment cycle. Both these factors combined to produce global recessionary conditions that also led to a collapse in share prices. (The impact on the share prices of some leading telecoms operators is shown in Fig. 1.2.) The downturn in Layer 2 immediately affected the companies in Layer 1 which produced the equipment needed by the network operators. Companies such as Lucent, Nortel, Ericsson, and Cisco were thrown into unprecedented crises, as will be shown shortly.[27]

Consequences for the Incumbent Telecoms Operators

One of the main consequences for the incumbent telecoms operators was downward pressure on revenue and profitability in their core businesses, particularly long-distance and international voice services. This pressure was the combined effect of increased competition due to low entry barriers and substantial new entry as well as technical advancement that significantly boosted capacity (and provided new innovations such as voice-over-IP, [Internet Protocol] the full effects of which have not yet been realized).

---

[27] The downturn in investment in telecoms equipment and software also had significant implications for the economy as a whole. In the US, the telecoms sector accounted for 12 per cent of all business spending on equipment and software in 2000. *Business Week*, 17 September 2001, p. 62.

The problem for the incumbents is that the bulk of their revenue and profitability still comes from these core businesses. It is true that these core businesses are increasingly being supplemented by new growth businesses, specifically mobile services (including voice and data), Internet-related, and fixed data services. But these new growth businesses have not, in the short to medium term, managed to compensate for the deterioration in the traditional core businesses. This is so for several reasons. First, the most important contribution from the new growth businesses launched by the incumbents has come from mobile services. However, while the revenue and profitability of mobile services have grown rapidly—and indeed have been the main bright spot in the incumbents' rather dismal portfolio of businesses—this growth has been insufficient to compensate for the downturn in the traditional core businesses. Second, although Internet-related businesses have generally been high-growth in terms of revenue, they have not been high-growth in terms of profitability. Furthermore, Internet-related businesses have generally only made relatively small contributions to the total revenue and profit of the incumbents. Third, the same has applied to data traffic which, generally, though a high growth area in terms of revenue, has not contributed significantly either to total revenue or profit.

How have some of the major global incumbents—such as AT&T, BT, Deutsche Telekom, France Telecom, and NTT—been affected by these trends? What have been their responses? These questions are answered in detail in Chapters 3–6.

Consequences for the New Entrant Telecoms Operators

The new entrant telecoms operators who developed and leased their own networks were eventually hit by the same destructive forces that were causing pain to the incumbents. This was particularly true for those new entrants that entered the long-distance and international voice and data business. Until around March 2000, for reasons that were analysed in detail earlier in this chapter, the share prices of the leading new entrants leaped ahead, both in absolute terms and relative to the incumbents. This is precisely what was predicted by the Consensual Vision.

However, from mid-2000 the gap in share price between the leading new entrants and the incumbents began rapidly to close. This too is shown in Fig. 1.2 for the cases of AT&T, WorldCom, and Qwest in the US and BT, Colt, and Energis in the UK. By the end of 2001 the gap had all but disappeared.

Why did the share price performance of the new entrants deteriorate so rapidly, refuting the hypotheses implicitly contained in the Consensual Vision? The answer to this question is that, paradoxically, it was the very success of the new entrants that undermined the ground on which not only they, but also the incumbents, stood. More specifically, the successful entry of the new entrants into the long-distance and international voice and data business, riding on the back of the new technology tiger, had the effect of undercutting the revenue growth and profitability not only of the incumbents, their arch-rivals, but also their own. Indeed, the performances of the incumbents and new entrants were the opposite sides of the same coin.

However, after mid-2000 the fortunes of the new entrants, which had looked so rosy in the light of the Consensual Vision, were now revealed to be threatened by serious weaknesses. Apart from the multiple factors already closely analysed in this chapter, the new entrants suffered from additional shortcomings. First, their businesses were exclusively in those areas which were most vulnerable to the combined effects of increased competition (through massive new entry)[28] and new technology, principally in the long-distance and international voice and data markets.

Second, their capital investment flows and their revenue flows were inadequately synchronized. Like their infamous cousins, the so-called dotcom companies, the new entrants, including the oldest and most advanced, were outperforming the general stock market more on the basis of promise than performance. Here was the contradiction on which their corporate strategies were based: the realization of their alleged superior competitiveness *vis-à-vis* the incumbents (proposed by the Consensual Vision) required significant investment both in networks and in supporting resources; however, given the *time* that it takes to build networks and supporting structures on the one hand, and to generate revenues on the other, these investments would have to be made *in advance* of the compensating revenue streams. While exuberant financial markets were willing to make good the gap between current investment and current revenue—as a result of optimistic expectations regarding the new entrants' future revenues—the clothes remained on the emperor. However, once these exuberant expectations changed to become pessimistic (a reflection of general changing stock market sentiment), the emperor's clothes quickly fell away to reveal a far more scrawny substance than had previously been believed.

The serious financial position of several major new entrants is illustrated in Table 1.2.

Table 1.2 speaks for itself, showing how seriously the financial position of some of the major new entrants had deteriorated by the end of the first quarter in 2001. Precisely the same fate was to befall some of the largest and best known of the new entrants—companies such as WorldCom, Qwest, Level 3, and Global Crossing. These were exactly the companies whose praise was sung, until early 2000, by Wall Street analysts such as Jack Grubman.

Of particular interest is the case of WorldCom, the oldest, largest, and perhaps best known of the US new entrants. In rosier days Grubman had argued that WorldCom was a 'must own'[29] share and the company the most promising

---

[28] Between 1996 and 1999 in the US alone 144 new telecoms companies went public, raising more than $25 billion.

[29] Jack B. Grubman and Sheri McMahon, 'WorldCom, Inc. Combination with MCI Creates the Only Legitimate Telecom Large-Cap Growth Stock', *Salomon Smith Barney Company Report*, 9 April 1998, p. 6.

Table 1.2. Financial position of some US new entrants, April 2001

Company	Debt ($ bn)	Cash ($ m)	Cashflow ($ bn)	Mkt.cap. at peak ($ bn)	Mkt.cap. Apr. 2001 ($ bn)
360 Networks	2.6	507	n.a.	18.8	1.60
Teligent	2.0	194	−1.0	6.0	0.04
Winstar	5.4	n.a.	n.a.	5.6	n.a.
XO Communications	7.3	1898	−2.0	24.4	1.40

Source: Financial Times, 30 April 2001.

of *all* telecoms network operators, with the best global footprint putting it in pole position to dominate the globalization process that was bound to follow soon.[30] Indeed, he argued that WorldCom should be placed in the same category, in terms of investment prospects, as the best performing US companies including Microsoft, Coca Cola, Walmart, Merck, and Disney.[31] Unfortunately, however, things were not to turn out for WorldCom the way Grubman and many others had predicted. Underlining the company's change in fortune, WorldCom's CEO, Bernard Ebbers, known more for his cowboy hat and boots and ebullient manner than for his humility, apologized publicly to his shareholders in 2001 for failing to deliver the shareholder value that he had promised them.

In Chapter 3 an analysis of WorldCom's performance is provided in comparison with that of AT&T, BT, and NTT. From this comparison it can be seen that an important advantage that the incumbents had over the new entrants came from their diversification into new growth areas. A particularly significant advantage for the incumbents—one not foreseen by either Ebbers or Grubman—came from their entry into mobile communications. This is discussed further in the next subsection but one.

Consequences for the Specialist Telecoms Equipment Suppliers

One of the strengths of the Layer Model shown in Table 1.1 is that it emphasizes the linkage and interdependence between layers 1 and 2 in the vertically specialized Telecoms Industry, namely the network operator layer and the telecoms equipment layer, respectively. Essentially, the network operators purchase most of their inputs (including a large part of their R&D) from the specialist telecoms

---

[30] 'We believe that there will be two to four global players who are fully integrated facilities-based providers of network services with WorldCom being the only fully integrated communications provider at the current time', Jack B. Grubman and Sheri McMahon, *op. cit.*, p. 6.

[31] Jack B. Grubman and Sheri McMahon, *op. cit.*, p. 5.

equipment suppliers. However, when the purchases of the network operators decline significantly, the equipment suppliers suffer from the decrease in demand.

From early 2000 the purchases of the network operators began to decrease significantly. This happened for a number of interrelated reasons. First, the period 1996–2000 saw huge increases in investment in network equipment as incumbent network operators and new entrants expanded their purchases and additional new entrants entered the market. It was shown earlier that these investments were fuelled by the Consensual Vision that predicted rapid increases in demand for capacity driven by the Internet. However, with the fall in the growth of revenue and profitability of the network operators (for reasons that were analysed earlier in this chapter) these investment rates in equipment could not be sustained. Second, we now know that the period 2000–2001 also saw a downturn in the economy-wide business cycle, beginning in the US and quickly spreading to Europe (with Japan continuing in a state of recession throughout the 1990s and into the new millennium). This meant cutbacks in telecoms expenditure by final business customers, further decreasing the derived demand for telecoms equipment. Third, the collapse of the dotcom companies from mid-2000 decreased significantly the demand for Internet-related services provided by network operators, such as web-hosting and data communications. Fourth, as the telecoms bust deepened, an increasing number of network operators collapsed—companies such as GTS, Viatel, Winstar, and Teligent in the US, and Atlantic Telecom in the UK. Furthermore, other leading companies such as Global Crossing and XO Communications teetered on the brink by the beginning of 2002.[32]

The net result of these trends was a significant decrease in demand for the equipment of the telecoms equipment suppliers—companies such as Lucent, Nortel, Cisco, NEC, Fujitsu, Siemens, and Alcatel—with dramatic consequences. Most dramatic was the turnaround in the fortunes of Lucent, the company that as part of the original Bell and AT&T systems had played such a key role not only in the development of the telecoms network in the US but also as a midwife giving birth to some of the major equipment companies in Europe and Japan.[33]

From the middle of 2000 Lucent's share price began a downward spiral when the company failed to meet Wall Street expectations of a 20 per cent growth in revenue, achieving only a 15 per cent growth. In October 2000, Richard McGinn,

---

[32] In 2001, in the US, twenty-eight publicly traded telecoms companies, each with more than $100 million of liabilities, sought protection from creditors in bankruptcy courts. *Business Week*, 28 January 2002, p. 57 and BankruptcyData.com

[33] Lucent was originally AT&T's Western Electric subsidiary. As Western Electric the company played a key role in the historical establishment and development of major companies such as ITT in the US, Nortel in Canada, what is now Alcatel in France, and NEC in Japan. For a detailed account of the historical role of Western Electric see M. Fransman, *Japan's Computer and Communications Industry* (1995).

Lucent's CEO, was dismissed by the company's board. For the first quarter of 2001 Lucent announced losses of over $1 billion and the Standard and Poor's credit rating agency downgraded the company's debt to junk status. In May 2001 an event occurred that would have been unthinkable a mere six months earlier, an event that sent shockwaves through the traditional telecoms equipment community: Alcatel, the French equipment company, made a bid to acquire Lucent (an offer that eventually collapsed when agreement could not be reached on who would control the new merged company).

The effect of the bust on Nortel, the global-leading Canadian telecoms equipment company, was equally dramatic. On 24 October 2000 Nortel's share price fell by 23 per cent in one day when the company failed to meet Wall Street's expected sales growth rate of 125 per cent (Nortel had achieved a mere 90 per cent). For the first quarter of 2001 Nortel announced a loss of $375 million and surprised many analysts when it declared that it was unable to give any guidance regarding its expected short-term financial performance. The company's fortunes declined with amazing rapidity. For the second quarter of 2001 Nortel announced a record loss for a quarter amounting to $19.5 billion (which included a write-off of goodwill in acquired companies bought during the telecoms boom).

Even Cisco, the high-flying company that dominated the world market for Internet routers, was hit by the downturn. Although CEO John Chambers assured analysts as late as December 2000 that he had 'never been more optimistic about the future of Cisco',[34] the company soon ran into difficulty. In May 2001 Cisco announced its first ever loss since the company went public in 1990 and Chambers was forced to admit that 'we are now in a valley much deeper than any of us anticipated'.[35]

Consequences for the Mobile Companies

Mobile Operators. Writing in April 1998, Jack Grubman stated that 'The one asset WorldCom does not own is wireless and we doubt that WorldCom will buy wireless assets anytime in the foreseeable future'. He saw this as something desirable and felt that it was unnecessary for WorldCom to develop its own mobile communications business:

Simply speaking, cellular/PCS [i.e. mobile communications] is not strategic for business customers since one cannot guarantee the integrity of a wireless network in the way that one can for voice or data...Frankly speaking, cellular remains largely a local exchange service and not one that is a critical part of the suite of voice, data, and IP network services that large business customers demand. Hence, we view WorldCom's and MCI's decisions not to pursue wireless assets as being beneficial for shareholder value without

---

[34] *Financial Times*, 6 April 2001, p. 20.   [35] *Financial Times*, 10 May 2001, p. 24.

impacting an iota the ability to provide global end-to-end connectivity [which Grubman saw as WorldCom's competitive strength].[36]

Arguably, however, it was CEO Bernard Ebbers' failure to lead WorldCom into mobile communications that constituted one of his most serious mistakes. It was a mistake, furthermore, that he later tried to correct when he made attempts to acquire Sprint PCS, the mobile business of Sprint, the third-largest US long-distance carrier. However, this attempt failed when it was blocked by the regulators.

Although the US was slower to develop mobile communications than both Europe and Japan (for reasons that are examined in Chapter 2), mobile provided a number of important advantages that greatly benefited WorldCom's competitors such as Verizon, Cingular (formed in 2000 by the merger of the wireless businesses of the former Baby Bells, SBC Communications and BellSouth), AT&T, and Vodafone. For one thing, Grubman and Ebbers, initially, seriously underestimated the rapidly growing demand for mobile telephony, not only in the US but also globally. For another, the mobile business provided the additional advantage that the limited number of licences issued by governments and regulators also served to limit entry and, therefore, competition. This made the mobile business the most lucrative in the entire telecoms industry from the end of the 1990s and into the first years of the new millennium.

The absence of a strong mobile business made WorldCom more vulnerable to the new forces of competition (analysed earlier in this chapter) that were unleashed in the company's core businesses of fixed long-distance and international voice and data. The effects of this competition started to show in the fourth quarter of 2000 when WorldCom's double digit revenue growth fell to around 8 per cent and the company announced that profits would be about half the $1.4 billion that analysts were expecting. As a result the company's share price fell by 20 per cent in a single day.

In July 1999, WorldCom had a market capitalization of $152 billion, making it the fourteenth most valuable company in the world. By October 2000, however, WorldCom's market capitalization had fallen to $52 billion. The contrast with Vodafone could hardly have been greater. As a 'single play' mobile company, Vodafone lacked all the advantages that Grubman saw in WorldCom. Both companies had the best global footprints in their respective businesses (fixed and mobile communications, respectively). However, Vodafone, which did not even appear in the list of the world's most valuable companies in July 1999, was, by October 2000, the most valuable company in Europe and the seventh most valuable in the world, with a market capitalization of $225 billion.

How important has mobile communications been for the diversified incumbent network operators such as AT&T, BT, Deutsche Telekom, France

[36] Jack B. Grubman and Sheri McMahon, 'WorldCom, Inc. Combination with MCI Creates the Only Legitimate Telecom Large-Cap Growth Stock', *Salomon Smith Barney Company Report*, 9 April 1998, p. 8.

Telecom, and NTT? The answer to this significant question is examined in detail in Chapters 3–6. It is worth noting here, however, that the benefit to the incumbents of their involvement in mobile communications has in some cases been significantly influenced by government policies relating to the auction of third-generation mobile licences. Particularly in the UK and Germany, where such auctions were first held, the prices of auctioned licences were extremely high, leading to a significant increase in the debt of some of the European incumbents. The impact on Deutsche Telekom and France Telecom is analysed in Chapters 5 and 6.[37]

Mobile Equipment Suppliers. The mobile equipment suppliers—such as Ericsson, Nokia, Motorola, and NEC—have also been hit by the Telecoms Bust just as have Lucent, Nortel, Cisco, and the other suppliers producing mainly for fixed networks. However, there have been special forces at work in the case of the mobile equipment suppliers as will now be shown.

To a significant extent, the mobile equipment suppliers have been negatively affected by the misfortunes of the incumbent telecoms operators that have already been analysed in this chapter. As the revenue growth and profitability of the incumbent telecoms operators deteriorated, and as sentiment in financial markets turned bearish with regard to telecoms companies, so less was spent by these operators on all telecoms equipment, including mobile equipment. Even though the incumbents' best performing businesses were in the mobile area, poor performance in other business areas together, in most cases, with high levels of indebtedness served to reduce expenditure on mobile equipment. This produced negative knock-on effects on the major companies supplying mobile equipment.

But there were also effects, specific to the mobile communications sector, that added to the woes of the mobile equipment suppliers. The first of these was the complicated transition from second-generation to third-generation mobile networks. To begin with, although it had been hoped that the third-generation would provide the first opportunity in the mobile sector for a global adoption of a single standard, this was not to be. Although Europe and Japan agreed on a single standard (W-CDMA), in the US there was a more patchy acceptance of this standard.

Furthermore, the transition from second to third generation took far longer than had been anticipated. In part, the culprit was the extremely complex nature of the mobile telecommunications system with its need for interoperability between different kinds of networks (including mobile networks of different generations). This made the task of developing mobile phones and other appliances capable of connecting to the different networks difficult. But the slower

---

[37] The pros and cons and impact of auctioned licences versus beauty-contents is too complex to go into here. For more information, however, see: http://www.TelecomVisions.com/current/auction.php.

than expected transition was also the result of performance-stretching innovations that improved capabilities provided by second-generation networks,
making them more competitive with third-generation networks at lower cost.[38]
In turn, this slowed the diffusion rate of third-generation networks.

Specific dynamics also impacted the area of mobile handsets. The slow
transition to third-generation mobile networks (and even to the so-called
2.5-generation networks, halfway between second and third-generation networks) meant that mobile handset makers could not count on completely new
demand for new kinds of handsets to continue fuelling the high growth rates
they had enjoyed with the rapid global diffusion of second-generation mobile
telephony. Instead, from 1999 to 2001 they had to rely on improving their
existing second-generation handsets through innovations (some of which, such
as different coloured changeable frontpieces, though technically trivial, had a
great impact on demand).

Unfortunately, for all but one of the main mobile handset producers, one of
their number had come to excel in the task of designing mobile handsets and
had accumulated distinctive competencies in this area that proved almost
impossible to imitate successfully.[39] This was the Finnish company, Nokia, that
by 2001 had come to dominate one-third of the global market for mobile
handsets, much to the consternation of its competitors. Falling handset sales
and profitability was a major factor that drove down sharply the share price of
companies such as Ericsson, Motorola, and Siemens, and even Nokia suffered a
severe fall despite its continuing dominance of global market share. While
competitors such as NEC-Matsushita hope to challenge Nokia with the advent
of third-generation networks, their ambitions remain to be tested.[40]

But perhaps the most important dynamic that will drive the future fortunes of
the mobile sector is consumer demand as mobile voice is increasingly replaced
as a source of revenue growth by mobile data services, including mobile
Internet. How great the demand for the latter services will be and how much
consumers will be prepared to pay constitutes a major cloud of uncertainty
hanging over the mobile sector as a whole.

[38] Innovations such as GPRS and EDGE provided some of the advantages over second-
generation networks that were promised by third-generation networks, but at significantly
lower cost. In the US a similar role is played for some mobile operators by CDMA 2000.

[39] The inimitability of Nokia's distinctive competencies in mobile handset design is
paralleled by the inimitability of Dell's manufacturing and more importantly marketing
processes in the area of PCs.

[40] NEC-Matsushita's joint venture involves a pooling of the two companies' R&D capabilities in the areas of telecoms equipment and consumer electronics, respectively. Combined, they will have about the same R&D expenditure as Nokia and they will both be
producing according to the same dominant world standard. The Japanese firms hope to benefit
from the extremely competitive mobile market in Japan, probably the most demanding in
the world.

Governments' Role

In this chapter, the important role played by governments has been left implicit. More specifically, little has been explicitly said about the crucial role played by governments as they liberalized telecoms services markets and attempted to increase competition between telecoms network operators through measures such as the regulation of interconnection charges and requirements. Clearly, many of these government interventions were a *sine qua non* for the processes of change that have been analysed in this chapter. However, since there is a huge specialized literature on telecoms regulation little more will be added here.

Could they and Should they have Known Better?

An examination of the flaws in the Consensual Vision, discussed earlier, suggests that more could have and should have been done in order to produce a more rigorous analysis of the dynamics of the Telecoms Industry in the Internet Age and the way in which the industry was changing. More specifically, more could have and should have been done to integrate the supply side of the industry with the projections that were being made regarding the demand side and to investigate the interactions between demand and supply. Furthermore, a more careful analysis of the barriers to entry confronting new entrants, could have and should have been undertaken, as is suggested in this book. This would have shown that it was unlikely that supernormal profits would continue to be made for very long. Such an analysis would have been enriched by an understanding of the process of vertical specialization in the industry, as is shown in detail here, a process which as analysts were beginning to understand, had already fundamentally transformed the computer industry.[41]

Had these kinds of analyses been undertaken, it is likely that the exuberance for the Telecoms Industry would have been tempered, at least for the medium to long term. However, even if such analyses had been undertaken, they would have been seriously constrained by the significant degree of interpretive ambiguity[42] surrounding some of the key parameters. One such parameter, for example, was future demand for bandwidth derived from future demand for bandwidth-intensive applications such as video-enabled web sites, video

---

[41] See Andy Grove (1996). The collapse in the profit margins on PCs, leading to the proposed merger of Hewlett Packard and Compaq in 2001, is causally similar to the fall in long-distance and international telecoms tariffs discussed in this book.

[42] Interpretive ambiguity may also be referred to as Knightian uncertainty after the Chicago economist Frank Knight who drew a sharp distinction between risk and uncertainty. According to Knight (1921), uncertainty exists when probability distributions cannot be derived.

conferencing and video-on-demand over the Internet. In the field of mobile communications similar interpretive ambiguity surrounded projections of likely demand for third-generation mobile services, a key determinant of the auction price that mobile operator bidders for 3G licences were prepared to pay. Similarly, the extent to which commercial transactions would migrate online, thus benefiting the dotcoms, and the speed at which this would happen, was bedevilled by interpretive ambiguity. As the industry moved further into the new millennium, estimates of these demand magnitudes became progressively more pessimistic. It is possible, however, that projections on the supply side, benefiting from somewhat more predictable technological trajectories (like Moore's Law), might have been more accurate.

As a result of interpretive ambiguity more rigorous analyses, of the type suggested here, would still not have yielded clear-cut answers regarding what would happen in the future and, therefore, what strategies and actions should be followed in the present. Also, it is precisely under such conditions of uncertainty that decision-makers tend to turn to cognitive frameworks such as the Consensual Vision discussed here in the hope that they will endow the decision-making process with more predictive powers. However, visions, even consensual ones, do not have the power to turn uncertainty into predictability. Furthermore, as the Telecoms Boom and Bust of 1996–2002 clearly shows, increasingly sophisticated theories and information-enhancing technologies also provide limited help. These, perhaps, are amongst the most important lessons to be learned.

Does this mean, returning to Grubman quoted at the beginning of this introduction, that all the groups that he referred to were 'culpable'? For the prosecution, it might be concluded that more could have been done to think the key issues through more rigorously. However, for the defence, whether this should have been done depends in part on the progress that would likely have resulted. And this progress would certainly have been severely constrained by the interpretive ambiguity that existed. At the very least, it might be concluded that more 'standing-back' would have been desirable, using the distance to look more critically at some of the key assumptions that were being made, explicitly and implicitly in the Consensual Vision, that in one form or another was being very widely employed to understand this industry. However, it would be wrong to imply that this would have resulted in the avoidance of mistakes and the making of correct decisions. Mistakes are an inevitable consequence of the irreducible uncertainty which always shrouds complex events such as those which occurred in the Telecoms Industry in the Internet Age from 1996 to 2002.

Having examined the Telecoms Boom and Bust, 1996–2002, in this chapter, we turn in the next to an examination of the forces driving the evolution of the Telecoms Industry.

Evolution of the Telecommunications Industry in the Internet Age

The Telecoms Industry's liberalization era began in the mid-1980s when Japan, the UK, and US took the first major steps towards opening their telecoms services markets to competition. In the mid-1990s, the Internet first emerged as a mass phenomenon. How did these radical changes affect the Telecoms Industry? What are the main forces that drive the evolution of this industry? How have the processes of knowledge-creation and use changed in the Telecoms Industry and with what consequences? When did mobile communications become important and why did its rate of diffusion vary so much in different parts of the developed world? These are the main questions analysed in this chapter.

INTRODUCTION

This chapter is about the process of evolution in the Telecoms Industry. More specifically, in this chapter, the processes of change are analysed as what is called the Old Telecoms Industry was transformed, first, into the New Telecoms Industry, and then under the influence of the Internet, into the Infocommunications Industry.

The demise of the Old Telecoms Industry began in the mid-1980s when, due to different combinations of political-economic circumstances, the monopoly of telecoms was ended in the US, UK, and Japan. By the late-1990s, with the agreement of the European Union to fully liberalize its telecoms markets and the similar agreement of the WTO, there was a widespread consensus that the liberalization of telecoms is essential. The roots of change that gave birth to the New Telecoms Industry in the early-1990s, however, as this chapter will show, were far more fundamental than the political and regulatory decisions that finally legitimized the changes. In the 1990s a new set of influences, that had begun thirty years earlier in an initially unrelated set of activities, brought about fundamental changes that further transformed the Telecoms Industry into the Infocommunications Industry. These influences came from the Internet based on its triad of core technologies: packet switching, Internet Protocol (IP), and the World Wide Web.

The Telecoms Industry and the Economics of Industrial Change

Despite important contributions made by earlier economists, such as Alfred Marshall and Joseph Schumpeter, it is reasonable to conclude that we do not yet have a comprehensive theory of the dynamics of industry, capable of explaining the process of change in specific industries, even though an increasingly rich body of knowledge is emerging in this area.[1] Indeed, it could even be conjectured that the search for such a comprehensive theory would inevitably be wrecked on the rocks of complexity and diversity. Under these circumstances a valuable purpose is served by detailed accounts of the evolution of particular industries which focus specifically on what Schumpeter called the 'engine'[2] of change.[3] What this engine consists of then becomes a key problem in the analysis. In the concluding section, several observations are made regarding the relevance of the Telecoms Industry for the economic analysis of the dynamics of industrial change.

THE TECHNOLOGICAL AND LEARNING REGIMES

A key part (though not the only part, see below) of the 'engine' driving change in the Telecoms Industry is the technological regime that exists in this

---

[1] For example, an important recent survey of economics literature on industrial dynamics by Dosi *et al.* (1997) concludes that 'certainly the gap between the richness of the histories [of the evolution of specific industries] and of their "appreciative" interpretations on the one hand and the theoretical models that are used to "explain" them on the other still appears quite large' (p. 20). Dosi *et al.* (1997), 'Industrial Structures and Dynamics: Evidence, Interpretations and Puzzles', *Industrial and Corporate Change*, 6(1): 3–24.

[2] 'The fundamental impulse that sets and keeps the capitalist *engine* in motion comes from the new consumers' goods, the new methods of production or transportation, the new markets, the new forms of industrial organization that capitalist enterprise creates'. Schumpeter (1943: 83) (emphasis added). As will be seen, it is proposed in this chapter that there are several engines driving the evolutionary process in addition to those proposed by Schumpeter.

[3] Such studies cannot be dismissed as inductive and descriptive for the simple reason that it is not possible to induce or describe the causes of industrial change. These causes do not emerge automatically or unambiguously from the information that we have on the complex process of industrial change. Rather, it is necessary to use, and constantly test, frameworks of interpretation in the attempt to come to grips with the tangle of causes that bring about industrial change. That alternative frameworks and conclusions are often possible, and accordingly have to be tested against the evidence, is part and parcel of the process of knowledge-creation in this area.

industry.[4] The *technological regime* is defined by the conditions under which technological knowledge is created—which determine the rate of technical change and the kinds of technologies that are created—and the opportunities and constraints that exist in the use of that knowledge. The technological regime, in turn, defines the *learning regime* that determines the kinds of learning paths and patterns in which the firms and other organizations involved in the industry will engage.

In order to understand the evolution of the Telecoms Industry—from the Old Telecoms Industry that existed until the mid-1980s, through the New Telecoms Industry, to the Infocommunications Industry—it is necessary to analyse the changing technological and learning regimes.

THE OLD TELECOMS INDUSTRY TO THE MID-1980s

Mapping the Old Telecoms Industry

A simplified map of the Old Telecoms Industry is provided in Table 2.1 in the form of a layer model.[5]

The Layers of the Old Telecoms Network. Layer 1 of the Old Telecoms Industry is the equipment layer where the network elements—such as switches and transmission systems—and customer premises equipment are produced that are combined to form and use telecoms networks. Until the 1970s these networks, shown in Table 2.1 in Layer 2, were primarily circuit-switched networks, where a dedicated circuit connects the sender and the recipient of information. In the 1970s, the first commercial packet-switched data networks made their appearance, although it was only in the 1980s, and even more so in the 1990s, when the Internet became widely adopted, that data communications and data

---

[4] The concept of technological regime used here is similar to that used in Nelson and Winter (1974, 1978, 1982) and in Winter (1984), although it is broader. For example, Winter (1984) distinguishes between two kinds of technological regimes: an entrepreneurial and a routinized technological regime. 'An entrepreneurial regime is one that is favorable to innovative entry and unfavorable to innovative activity by established firms; a routinized regime is one in which the conditions are the other way round' (p. 297). Audretsch (1997) uses the concept in the same way as Nelson and Winter. Malerba and Orsenigo (1990, 1993) and Dosi *et al.* (1997) provide a further elaboration of the concept of technological regime. More specifically, in the latter reference, it is stated that 'a technological regime is a particular combination of some fundamental properties of technologies: opportunity and appropriability conditions; degrees of cumulativeness of technological knowledge; and characteristics of the relevant knowledge base' (p. 94).

[5] Further details on the layer structure of the Telecoms Industry are to be found in http://www.TelecomVisions.com. I would like to thank Sadahiko Kano for the role that he played in developing the Layer Model that is analysed in this interactive web site. The Layer Model used in this chapter was also used in Chapter 1.

Table 2.1. Layers of the Old Telecoms Industry

Layer 3: Services layer
(voice, fax, 0800 services)
Layer 2: Network layer
(circuit-switched network)
Layer 1: Equipment layer
(switches, transmission systems, customer premises
  equipment)

---

services became financially important. As shown in Layer 3, the main services in the latter stages of the Old Telecoms Industry were voice, fax, and enhanced services such as toll-free 0800 services. Later in this chapter, we shall return to the layer model in order to examine the impact of the Internet on what will be referred to as the Infocommunications Industry.

Monopoly, Vertical Integration, and Quasi-vertical Integration

In the days of the Old Telecoms Industry the conventional wisdom was that telecoms were an example of 'natural monopoly', that is, due to increasing returns to scale, telecoms services could only be provided efficiently by a monopolist provider. Accordingly, in most industrialized countries (Finland being a notable exception), Layer 2, the network layer, was dominated by a monopolist network operator.

The natural monopoly hypothesis, however, was not, by and large, extended to Layer 1, the equipment layer. In different countries, the production of telecoms equipment was organized in different ways. At the one extreme was the US, where a pattern of vertical integration emerged almost from the birth of the Telecoms Industry.

In the US:

... from the time that Alexander Graham Bell cooperated with instrument-maker Thomas Watson in producing the first telephone sets, it was the same organization that both developed the telecommunications network and developed and manufactured the equipment that it required. This pattern was firmly established in 1880, when the American Bell Telephone Company purchased Western Union's telephone supplying subsidiary, the Western Electric Company of Chicago. According to an 1882 agreement, American Bell restricted itself to purchasing all its telephone equipment from Western Electric, while the latter agreed to limit its activities to supplying American Bell and its licensees.[6]

This vertical integration of network operation and equipment production in AT&T continued until the company's voluntary trivestiture in September 1995

---

[6] See Fransman (1995: 24).

into one company providing telecoms services, the new AT&T, one providing equipment, Lucent, and one providing computers and computer services, essentially the former NCR that had been acquired in a hostile takeover by AT&T in 1993.

At the other extreme were the smaller industrialized countries (Sweden with Ericsson being a notable exception) and most of the developing countries where the national monopoly telecoms carriers procured their telecoms equipment from the handful of specialist telecoms equipment suppliers from other countries who competed in world markets for telecoms equipment. While being profitably locked into long-term relationships with the monopoly network operators in their home country, these specialist technology suppliers competed vigorously in the rest of the world, where telecoms equipment markets were not similarly locked up by competing suppliers with privileged supply relationships with the national operator.

In the middle were the large industrialized countries with domestic markets that were sufficiently large to support domestic telecoms equipment production. In this category of countries, however, the economic organization of Layers 1 and 2—the equipment and network layers—differed significantly.

In Japan, for example, from the late-nineteenth century, when the Ministry of Communications took responsibility for the development of the new telecoms infrastructure, the decision was made for several competing companies to produce the telecoms equipment required for the Japanese telecoms network. In this way, a family of four specialist telecoms equipment suppliers emerged to supply the Ministry under a form of economic organization that has been referred to as 'controlled competition'.[7] The lead company was NEC, founded in 1899 as a majority-owned subsidiary of Western Electric, the equipment supplying subsidiary of AT&T. The other three members of the family were Fujitsu, which had an ownership link with Siemens, Hitachi, the only independent Japanese telecoms equipment supplier, and Oki. This family of suppliers continued to supply NTT, the national monopoly operator, which was separated from the Ministry in 1952 as an independent state-owned company.

In Britain, the Post Office which had responsibility for telecoms (later separated as BT) also cooperated closely with national telecoms suppliers that included GEC, Plessey, and STC (a subsidiary of the US firm ITT). However, the British experience with a long-term obligational cooperative relationship between network operator and equipment supplier was not nearly as happy as in the Japanese case. By the time BT was privatized in 1984 and liberalization began in the British telecoms market, the relationship between BT and its privileged national suppliers had already started to crumble.[8]

---

[7] For a detailed account of the origins of the Japanese system of 'controlled competition' in telecoms, see Fransman (1995: 27–41).

[8] For a detailed account of the less-than-happy British attempt at cooperative research and development in the case of telecoms switches, involving the monopoly operator and its family of suppliers, see Fransman (1995: 89–97).

In France and Germany, the monopoly network operator—to become France Telecom and Deutsche Telekom, respectively—also cooperated closely with national equipment suppliers. In France, a complex process of government-inspired reorganizations and mergers, largely between the subsidiaries of the American company ITT and French electronics companies, resulted in the birth of the major French specialist telecoms equipment company, Alcatel.[9] In Germany, it was the electrical and electronics company, Siemens, that immediately became the major national equipment supplier although the German government also procured equipment from non-German suppliers.

To conclude, in Japan, Britain, France, and Germany, a pattern of close, long-term, obligational cooperation emerged as the dominant form of economic organization between the monopoly operator in Layer 2 and the specialist technology suppliers in Layer 1. Although the degree of competition between the national suppliers in their national market differed—with Japan insisting on the strongest degree of competition—it is reasonable to characterize the dominant form of economic organization in these countries as one of quasi-vertical integration.

The Technological Regime in the Old Telecoms Industry

Earlier we stated that the technological regime is defined by the conditions under which technological knowledge is created and the opportunities and constraints that exist in the use of that knowledge. What were the conditions under which technological knowledge was created in the Old Telecoms Industry? To answer this question, it is necessary to understand the *innovation system* that existed in the Old Telecoms Industry.

Essentially, in the Old Telecoms Industry, the 'engine of innovation' was located in the central research laboratories of the monopoly telecoms operators such as AT&T's Bell Laboratories, BT's Martlesham Laboratories, France Telecoms's CNET Laboratories, or NTT's Electrical Engineering Laboratories. It was in these laboratories that the research was done that would lead eventually to the next generations of switches, transmissions systems, and other telecoms equipment that would improve telecoms services. Many of the key technologies still driving the Infocommunications Industry began life in these central research laboratories—such as the transistor, the laser, the design of cellular mobile systems, and the software language C that all emanated from Bell Laboratories.

Typically, after the central research laboratory did the initial research, and developed and tested the initial prototypes, the task of further development and mass manufacture was handed on to the specialist equipment suppliers—such

---

[9] For a detailed account of the birth of Alcatel, see Fransman (1995: 87–9).

as Western Electric in AT&T's case or NEC, Fujitsu, Hitachi, and Oki for NTT. Over time, however, these specialist equipment suppliers increased their own R&D capabilities with the result that they eventually took over many of the innovative tasks that in the Old Telecoms Industry were performed by the central research laboratories of the monopoly telecoms operators. As we shall see, this created a fundamental change in the technological and learning regimes that was to profoundly change the entire structure of the Telecoms Industry.

How efficient was the innovation system of the Old Telecoms Industry? In answering this question, it has to be said that despite the national monopoly position enjoyed by the telecoms operator, and the small number of privileged specialist suppliers who created the equipment for the telecoms network, the innovation system worked remarkably well. Evidence for this conclusion comes from the impressive stream of both radical and incremental innovations that emerged from the central research laboratories of the monopoly operators. One performance benchmark comes from the fact that in the US, the price of a local phone call remained constant in money terms for about one hundred years.

The paradox of a rapid rate of radical and incremental innovation from a system dominated by a monopolist supplied by no more than four privileged suppliers is to be explained in terms of the non-market incentives for innovation that nevertheless existed in the Old Telecoms Industry. The first of these non-market incentives came in the form of the 'cooperative competition' that existed between national systems to be the first to introduce the next generation of technologies and services. One example is the races that took place to develop the next generation of telecoms switches, races that were nonetheless punctuated by the formal and informal sharing of information through institutions such as regular international switching conferences that brought together the world's best.[10] The second major non-market incentive came in the form of political incentives and pressures to improve telecoms services for both residential and business users, who together constituted the bulk of the population and, therefore, wielded political muscle.

However, despite the impressive innovative performance of the Old Telecoms Industry, there were several characteristics of the innovation system in this industry that constituted important constraints on the innovation process, constraints that can be highlighted by a comparative analysis of the new innovation process in the New Telecoms Industry that replaced it (examined in more detail below).

The characteristics of the innovation system in the Old Telecoms Industry are summarized in Table 2.2.

To begin with, the innovation system in the Old Telecoms Industry was closed in the sense that only the monopoly telecoms operator and its chosen circle of specialist equipment suppliers were given access to the telecoms

---

[10] For a detailed study of the development of telecoms switches in the US, Japan, and Europe and for the development of optical fibre in the US, Japan, and the UK, see Fransman (1995).

Table 2.2. Characteristics of the innovation system in the
Old Telecoms Industry

Closed innovation system
High entry barriers
Few innovators
Fragmented knowledge base
Medium-powered incentives
Slow, sequential, innovation process: Research–prototype–
    trials–cutover

network and were able to make innovations for it. This implied that there were
extremely high, even prohibitive, barriers to entry into this innovation system.
Accordingly, there were very few innovators. Furthermore, the knowledge base
underpinning global telecoms was fragmented. Each national telecoms network
had its own specific designs and technologies. For example, central office switches designed for the Japanese market could not, without the cost of considerable modification, be immediately deployed in Europe or the US. The same
was true for switches designed for the European or US markets. Under this
regime, incentives were medium-powered since the market size for products
designed for specific markets was relatively small.

Finally, the innovation process itself was slow and sequential. In the circuit-
switched network, it was necessary that switching and transmission equipment
worked with extremely high degrees of reliability. Equipment failure could lead
to entire networks being shut down. Typically, the innovation process began
with research being done in the central research laboratories of the leading
monopoly operators. This research eventually resulted, in the second stage, in a
prototype being developed. In the third stage, the prototype was exhaustively
tested, first in the laboratory and then, after its further development and
manufacture by the specialist equipment supplier(s), in field trials. Finally, after
the reliability of the prototype had been adequately demonstrated, the equipment was 'cutover' into the network. Although this innovation process 'worked'
in the Old Telecoms Industry, its drawbacks became apparent, as we shall see
below, in the light of the fundamentally different process that replaced it in the
New Telecoms Industry.

The Learning Regime in the Old Telecoms Industry

The technological regime structures the learning regime. Essentially, the
learning process involved the monopoly network operator (located in Layer 2)
learning how to run and improve its telecoms network which provided the
platform for the services that it provided. Being a monopolist, the 'selection
environment' of the network operator excluded pressures and incentives to

compete in telecoms services markets. However, as noted in the previous section, there were pressures and incentives to improve both the network and services emanating from domestic political processes as well as rivalry between national systems to rapidly introduce new technologies and services. The latter pressures and incentives shaped the learning process.

In the Old Telecoms Industry, the monopoly network operator was both a user and an innovator of telecoms technologies and equipment. However, there was a division of labour with the network operator concentrating on research (including fundamental and long-term research) and design, while its selected equipment supplier(s) (located in Layer 1) specialized in development and mass manufacture of equipment. As a sophisticated user and innovator of telecoms technologies and equipment, the monopoly network operator was well placed to learn-by-using and by experiencing opportunities for further improvements by running the network.

However, in developing telecoms technologies and equipment for mass manufacture, the specialist equipment suppliers were also undergoing a learning process that enabled them over time to move into the upstream parts of the innovation process, namely into increasingly research- and design-intensive activities. Moreover, while they may have enjoyed sheltered access to the procurement book of the monopoly network operator in their own domestic market, in order to grow, they were forced to enter and compete in foreign markets, particularly third-country markets where there were not similarly sheltered equipment suppliers producing for the domestic monopoly network operator. Competitive pressures and incentives in these markets were an important stimulus for learning by the specialist equipment suppliers.

The learning process was also structured by technological paradigms and trajectories. For example, the transition to each new generation of central office switch—from electro-mechanical to space-division electronic to digital time-division to asynchronous transfer mode (ATM) switches—was accompanied by considerable controversy over the question of whether improvements in the old paradigm would make the new paradigm irrelevant.[11] Overall, however, it was the circuit-switched paradigm that shaped thinking and learning about how to achieve improvements. The deeply ingrained influence of the circuit-switched paradigm on the part of telecoms researchers and engineers is apparent in the reception that Vinton Cerf and his colleagues (who were amongst the original founders of the Internet) encountered from Bell Labs researchers when details about the packet switching that they were developing became known. In Cerf's words: 'The packet-switching network was so counter-culture that a lot of people thought it was really stupid. The AT&T guys thought we were all beside ourselves; they didn't think that interactive computing was a move forward at all'.[12]

---

[11] See Fransman (1995: ch. 3) for further details.

[12] *Internet*, 'Is there a future for the Net? David Pitchford finds out from the man who invented it, Vinton Cerf', 19 June 1996, p. 75.

Coming from a computer culture, the developers of the ARPANET, the forerunner of the Internet, were not inclined, as were their telecoms counterparts, to see the problem of communications through the paradigmatic prism of circuit switching (as will be seen in more detail later).

THE TRANSITION TELECOMS INDUSTRY

In the mid-1980s, for different political-economic reasons, Japan, the US, and UK decided to end the monopolies of their monopoly network operators. The result was the birth of the original new entrants.

The Birth of the Original New Entrants

The introduction of liberalization and competition in these three countries, however, was at first cautious and tentative. In Japan, three long-distance competitors were given regulatory permission to compete with NTT, namely DDI, Japan Telecom, and Teleway Japan. NTT was only partially-privatized, with the Japanese government continuing to own approximately two-thirds of the company. The UK government, on the other hand, soon sold off the majority of BT's shares but began the process of liberalization with a period of duopoly with Mercury, a subsidiary of Cable and Wireless, as the sole competitor to BT. In the US, AT&T was divested with the new AT&T inheriting the former company's long-lines business (i.e. long distance) while seven regional companies—the Baby Bells—retained the de facto monopoly of local telecoms services in their regions. MCI and Sprint were the two long-distance companies allowed to compete with AT&T.

Original New Entrants and Specialist Technology Suppliers

Although liberalizing regulatory regimes provided a necessary condition for rapid and successful entry by the original new entrants, they were not sufficient. Equally important were low technological barriers to entry into the telecoms services markets (in Layer 2) created by the existence of specialist telecoms equipment suppliers. These specialist technology suppliers provided the 'black-boxed' technologies that the Original New Entrants needed to construct and run their own networks. Unlike their counterparts in other industries such as pharmaceuticals or semiconductors, where substantial in-house technological capabilities were necessary in order to enter and compete, the Original New Entrants in the Telecoms Industry were able to turn to specialist technology suppliers for most of their technology requirements. Without the knowledge-acquisition and learning processes that the specialist technology suppliers underwent in the Old Telecoms Industry, would-be original new entrants would have faced formidable technological barriers to entry.

From the point of view of the specialist technology suppliers, liberalization created new markets for their accumulating knowledge and competencies. An important example is Nortel that seized the new opportunities presented by liberalization with both hands. Nortel was originally established in 1895 as the subsidiary of Bell Canada. From 1906 to 1962, AT&T's equipment subsidiary, Western Electric, held a minority stake in Nortel, a stake that was gradually sold to Bell Canada. In 1971, Bell Canada, that bought most of its equipment from Nortel, established a joint R&D subsidiary with Nortel called BNR. However, in order to grow, Nortel, from the late 1970s, made strenuous efforts to enter export markets. In these attempts, the company was considerably aided by its pioneering success in developing one of the first small digital central office switches, the DMS 10. Beating AT&T into this segment of the switching market, Nortel was able to gain a foothold in the US, its first major breakthrough outside Canada.[13]

In MCI, the main long-distance competitor to AT&T, Nortel found an important ally. As a competitor to AT&T, MCI, like many of its original new entrant counterparts in other parts of the world,[14] did not want to depend on the same specialist equipment suppliers that supplied the incumbent. This provided Nortel with the opening it was seeking. Furthermore, able to rely on Nortel and other specialist technology suppliers for the technology and equipment it needed, MCI also decided that it did not need to replicate similar R&D capabilities that AT&T had in its Bell Laboratories.[15] With further stages of liberalization in the US and Europe, Nortel quickly became the main technology supplier to the 'new new entrants' as will be shown below.

The Conservatism of the Original New Entrants

Perhaps because the selection environment into which they were born was characterized by only partial competition, the original new entrants in the US and UK were conservative rather than radical in their competition with their incumbent, tending largely to imitate them while only slightly underpricing them. Soon, however, the original new entrants would be overshadowed by the

---

[13] Further details of Nortel success are to be found in Fransman (1995: 55–61), including an explanation of why Nortel managed to beat AT&T in this switch market segment.

[14] DDI, the main competitor to NTT, similarly refused to procure its equipment from NTT's family of specialist equipment suppliers, NEC, Fujitsu, Hitachi, and Oki.

[15] Indeed, MCI took particular pride in the fact that, with access to specialist technology suppliers it did not need to make the same expenditures on R&D as AT&T. While AT&T was spending more than 7 per cent of its sales on R&D, MCI's expenditures were so negligible that they were not even reported. (Author's interviews with Serge Wernikoff, Senior Vice President and Board Member of MCI, 1993, 1994.)

new breed of 'new new entrants' whose entry signalled the emergence of the New Telecoms Industry.

THE NEW TELECOMS INDUSTRY FROM THE EARLY-1990s

Enter the New New Entrants

In retrospect, from the vantage point of the late-1990s, it was clear that a qualitative change had occurred in the Telecoms Industry in the early-1990s, signifying the birth of the New Telecoms Industry. The most evident sign of qualitative change was the rise of the new new entrants that quickly eclipsed the original new entrants and went on to pose the most significant threat to the incumbents.

Most dramatic was the rise of WorldCom, a company that was born in 1984 in the inauspicious location of Hattiesburg, Mississippi, and began life as a reseller of the newly divested AT&T's capacity before making the key strategic decision to become a facilities-based operator. By the end of the millennium, not only had WorldCom capped a string of mergers and acquisitions with the takeover of MCI, the main long-distance competitor to AT&T, it also boasted the world's best global telecoms network making this company the most serious threat to the Big Five Incumbents—AT&T, BT, Deutsche Telecom, France Telecom, and NTT. Although they emerged later than WorldCom and were not as large in terms of revenue and market capitalization, several other new new entrants replicated essentially the same growth process. These companies included Qwest, Level 3, Global Crossing, Williams, and Viatel.

In Europe, new new entrants made a similar, though somewhat later, dramatic entry. These new players in Europe included telecoms operators such as City of London Telecommunications (COLT), Energis (the English telecoms subsidiary of the regional electricity supplier), and Mannesmann (the German industrial engineering company that transformed itself into both a fixed-line and mobile operator).

Unlike the original new entrants, the new new entrants were far more aggressive in their competition. This was seen both in their activities in the market for corporate control as well as in global telecoms markets. In the late 1990s, for example, WorldCom acquired MCI (as already noted), Qwest acquired US West (the smallest Baby Bell), and Global Crossing acquired Frontier. Furthermore, all the US new new entrants began to construct their own global networks even before their US networks were fully completed, typically beginning in the UK, the most liberalized of the European markets, and then moving on to construct pan-European networks. (This raises an important puzzle about the characteristics of the US telecoms system that explain why the most aggressive telecoms operators, in terms of the global expansion, are

American.)[16] It was only in Japan that by the end of the millennium, new new entrants had not displaced the original new entrants.[17]

Why were the new new entrants able to make such a dramatic entry and why were they able to be so successful so rapidly? A large part of the answer to this important question lies in the changes that occurred in the technological regime, changes that began already in the late stages of the Old Telecoms Industry.

The Technological Regime in the New Telecoms Industry

By the mid-1990s, a decisive process of vertical specialization had occurred between Layers 1 and 2 (see Table 2.1) in the Telecoms Industry. As noted earlier, in the Old Telecoms Industry, the R&D engine was largely located in the central research laboratories of the monopoly network operators such as AT&T, BT, and NTT, with the specialist equipment suppliers being largely relegated to playing the role of dependent developers and manufacturers. By the end of 1995, however, this situation had changed dramatically with the now incumbent network operators making the decision to leave more and more of the R&D related to the network and its elements to the specialist equipment suppliers. At the same time, the incumbents decided to open their procurement, agreeing to buy from new suppliers in addition to their traditional suppliers.

A snapshot of the technological regime in transition, as it moved from the Old to the New Telecoms Industry, is provided in an article published in 1994 that compared AT&T, BT, and NTT in terms of their visions, strategies, competencies, and R&D.[18] This article amongst other things showed that these three companies had made fundamentally different organizational decisions regarding their technological competencies and their procurement of telecoms equipment. While AT&T resorted to vertical integration, procuring the bulk of its equipment from its own equipment business, BT, at the other extreme, had decided increasingly to resort to the market for its equipment needs. In the middle, NTT relied on its own home-grown form of organization, namely controlled competition involving a closed family of four suppliers, for its equipment supplies. The different choices made by these three companies were reflected in their markedly different R&D intensities. While AT&T spent about 7 per cent of its sales on R&D, the figure was around 4.5 per cent for NTT, and about 1.9 per cent for BT.

---

[16] As will be made clear later, this generalization, while true for the fixed-line operators is untrue for the mobile operators. In the mobile field, for reasons that will be analysed below, it was European and to a lesser extent Japanese companies that dominated globally.

[17] This is a further puzzle that also requires explanation in terms of the characteristics of the Japanese telecoms system. Unfortunately, however, there is insufficient space here to explore this puzzle.   [18] See Fransman (1994), reprinted in Fransman (1999).

By the end of 1995, however, a decisive process of convergence began to take place among these three companies. AT&T underwent the biggest change when, in September 1995, the company voluntarily trivested itself, spinning off its equipment business in the form of Lucent, and NCR, the computer company that it had acquired in a hostile takeover in 1993. BT continued along the path that it had charted from the early 1990s, after its unhappy experiences in jointly developing its own System X digital switch, and left more and more of its network needs to the market. NTT, to begin with, to some extent under trade-related pressure from the US but later increasingly responding to the new opportunities provided by vertical specialization, began to procure a greater proportion of its equipment from outside its traditional family of suppliers and to shift its R&D away from network-related areas, and more towards service-related innovation.

The Changing Location of R&D in the New Telecoms Industry.[19] One of the best indicators of change in technological regime as the Old Telecoms Industry gave way to the New Telecoms Industry is provided by data on the changing location of R&D. This data is summarized in Table 2.3.

Several characteristics of the technological regime in the New Telecoms Industry are evident from Table 2.3. The first characteristic is that the incumbent network operators—represented in Table 2.3 by NTT, BT, and AT&T—are not particularly R&D-intensive. Indeed, the bottom cell in Table 2.3 shows that the incumbents are less R&D-intensive than the average in industries that are not normally thought of as 'high-tech' industries, namely the vehicle, leisure and hotel, building materials, and brewery industries.

The second characteristic is that the new new entrants—represented here by WorldCom, Qwest, Level 3, and Global Crossing—are even less R&D-intensive than the incumbents, doing virtually no in-house R&D. The reason, as already mentioned in this chapter, is that the new new entrants have made the strategic decision to outsource almost all of their R&D requirements to the specialist technology suppliers. The reasoning given by Energis, one of the major new new entrant rivals to BT in the UK, for not doing its own R&D is typical for all the new new entrants: 'Energis' policy is not to undertake significant research and development, but to utilize technology developed by its suppliers and, as a result, Energis has spent an immaterial amount on research and development . . . '.[20]

As a substitute for its own internal R&D, Energis has forged a highly satisfactory relationship with Nortel as its main technology supplier, a relationship that has also involved the subcontracting of specific R&D projects.[21] Similarly, Qwest, the US new new entrant that in 1999 acquired the smallest of the Baby Bells, US West, also undertakes virtually no internal R&D. In its 1998 Annual

---

[19] In Chapter 8, the changing role of R&D in the Telecoms Industry is analysed in more detail.
[20] Energis, *Global Offering*, Form 20-F, p. 6.
[21] Author's interview with Mike Grabiner, CEO of Energis, 1999.

Report, Qwest states that its 'R&D costs incurred in the normal course of business are...$27.7 million...in 1998'. This compares with the company's total revenue for 1998 of $2,242.7 million, which makes Qwest's R&D expenditure a mere 0.012 per cent of total revenue.[22] As Qwest freely admits, 'We built our network with state-of-the-art technology and through alliances with companies like Cisco, Nortel and Ascend. In 1998 we continued to join with the best in the business to take the power of network applications to more markets in new ways with advanced products'.[23]

The third characteristic is that R&D-intensive activities, mainly relating to the elements that go into networks, have moved decisively into the specialist technology suppliers, represented in Table 2.3 by Cisco, Ericsson, Nortel, Lucent, and Nokia. These specialist technology suppliers are some six times as R&D-intensive as the incumbents, AT&T, BT, and NTT. Furthermore, their R&D-intensity is comparable to that of the pharmaceutical companies, acknowledged to be amongst the most R&D-intensive of all sectors. This is evident from the figures provided in Table 2.3 for the pharmaceutical companies Roche, Glaxo Wellcome, and SmithKline Beecham. And the five specialist technology suppliers represented in Table 2.3 represent just the tip of the iceberg. In addition to the other large telecoms equipment suppliers such as Alcatel, Siemens, NEC, and Fujitsu are the huge number of R&D-intensive medium-sized companies that are supplying significant telecoms technologies in numerous niches.

It may be concluded, therefore, that while in the Old Telecoms Industry the 'innovative engine' was located largely in the central research laboratories of the monopoly network operators, in the New Telecoms Industry, the 'R&D engine' has moved decisively into the specialist technology suppliers. This provides one key indicator of the extent of the process of vertical specialization between Layers 1 and 2 (see Table 2.1) in the New Telecoms Industry.

A significant word of caution, however, is necessary regarding this conclusion. This arises because it is important not to confuse R&D with innovation. Firms with low R&D intensity may nevertheless be highly innovative, and their innovativeness may lead to competitiveness and high profitability. One example is MCI's Family and Friends billing innovation that allowed the company to offer preferential tariffs on several frequently called numbers, and for a while gave the company a significant advantage over its rival, AT&T. Although not falling under the conventional classification of R&D, this innovation is indicative of the kinds of advances that may be made that do not involve R&D. We shall return later to the importance of innovation in the discussion of the learning regime in the New Telecoms Industry.

Specialist Technology Suppliers and Low Technological Barriers to Entry. One part of the answer to the puzzle of why the new new entrants were able to enter the Telecoms Industry so rapidly and so successfully is now, in the light of the

---

[22] Qwest, *Annual Report*, p. 33 and p. 28, respectively.   [23] Qwest, *Annual Report 1998*, p. 11.

Table 2.3. The location of R&D in the New
Telecoms Industry, 1999

Firm/industry	R&D % sales
NTT	3.7
BT	1.9
AT&T	1.6
Cisco	18.7
Ericsson	14.5
Nortel	13.9
Lucent	11.5
Nokia	10.4
WorldCom	~0
Qwest	~0
Level 3	~0
Global Crossing	~0
Roche	15.5
Glaxo Wellcome	14.4
SmithKline Beecham	10.8
Vehicle industry	4.2
Leisure & hotel industry	3.2
Building materials industry	3.0
Brewery industry	2.3

Source: *Financial Times*, R&D Scoreboard,
1999, and author's calculations from company
reports.

analysis of the changing technological regime, apparent. Being able to rely
on highly competitive markets for technology supplied by a host of rivalrous
specialist technology suppliers, the new new entrant operators have faced
extremely low technological barriers to entry.

This also explains the apparent paradox of firms that to begin with knew
virtually nothing about telecommunications becoming telecoms operators.
Examples include Bernard Ebbers who was one of the founders of WorldCom
but had a background as a football coach and motel operator; COLT one of the
main challengers to BT in the UK, which was established by Fidelity, the largest
mutual fund in the US (that earlier had established Teleport, a competitive local
exchange carrier in New York and Boston that was later sold to AT&T);[24] Qwest,
established by Philip Anschutz, a billionaire with a background in ranching, oil,
and railroads;[25] and Mannesmann, Olivetti, and Energis that came from backgrounds in industrial engineering, electronics, and electricity, respectively.

[24] COLT is the subject of Chapter 7.
[25] Qwest is discussed in more detail in Chapter 6.

With such low technological barriers to entry, the result has been a highly competitive market for network services (in Layer 2, see Table 2.1).[26] However, the contribution of specialist technology suppliers was not confined to the supply of technology. Also important are the human resources that these suppliers provided, through the operations of the labour market, to both the original new entrants and the new new entrants. An examination of the backgrounds of the leading executives in the major original new entrant and new new entrant telecoms companies readily reveals the importance of recruitment from specialist technology suppliers (as well as from the incumbents). Furthermore, there are key examples where specialist technology suppliers, like Nortel and Ericsson, have also provided their new new entrant customers with finance, that in some cases has played a major role in facilitating the growth of these companies.

Also central has been the role played by capital markets. In some cases, capital markets have interacted with labour markets to support the entry and growth of new entrants, for example, where share option schemes, with the expectation of significant future increases in share values, have provided an important incentive for people possessing key knowledge to move from specialist technology suppliers, or incumbents, to new entrants. The role of capital markets is analysed in more detail in a separate section below.

The Learning Regime in the New Telecoms Industry

As noted earlier, the technological regime structures the learning regime. What are the main learning processes that occur in the New Telecoms Industry?

Learning by Network Operators.[27] Table 2.3 above suggests that it is necessary, in examining the learning process in the New Telecoms Industry, to distinguish between the R&D-performing incumbent network operators—such as AT&T, BT, Deutsche Telecom, France Telecom, and NTT—and the R&D-less new new entrants—such as WorldCom, Qwest, COLT, and Energis. In the former companies, the learning process includes activities undertaken in organizationally distinct R&D laboratories, whether they are attached to business units (where the bulk of the incumbents' R&D resources are located) or situated in central research facilities. In the latter companies, as we have seen, R&D learning is

[26] A further important contributor to competition in Layer 2 has been the proliferation of competing network technologies, a topic that is taken up later in the section dealing with different forms of competition in the New Telecoms Industry.

[27] Attention in this section is confined to those telecoms companies that own and operate their own networks, that is, to the larger companies in Layer 2 (see Table 2.1). In a very different position are facilities-less telecoms service providers—such as call-back providers, resellers, or facilities-less Internet Service Providers. Since these kinds of telecoms companies buy-in the services of network operators, their learning processes are fundamentally different from those of facilities-based telecoms companies. We shall return to facilities-less telecoms service providers in our discussion below of the layer model for the Infocommunications Industry shown in Table 2.7. See TelecomVisions.com for further details on the facilities-less service providers.

entirely outsourced to specialist technology suppliers and the fruits of this learning bought-in in the form of tangible assets such as telecoms equipment or intangible knowledge such as occurs when technological advice is given.

However, all network operators participate in the division of labour that follows from the process of vertical specialization between Layers 1 and 2 (see Table 2.1). In particular, all network operators are heavily dependent on purchases from specialist technology suppliers of technology in the form of network-related equipment and associated systems such as billing and IT management software. This dependence is even greater in the case of the new new entrants who, as we have seen, do not undertake their own R&D. Nevertheless, since all of the network operators are very dependent on specialist technology suppliers, much of their learning takes the form of learning to *use*, rather than learning to *produce*, technology.

New New Operators as F4 Firms. A further distinguishing feature of the new operators, in contrast to the incumbents, is their ability to focus on a subset of market segments. This ability has resulted largely from the absence of universal service obligations (even though regulatory regimes sometimes require new operators to contribute financially to universal service). This has meant that new operators have been able to choose particular market segments—such as multinational business, large domestic business, small and medium-sized business, or residential customers—and focus their learning processes on the chosen segments. In turn, their focus on particular customer segments and their absence of universal service obligations has allowed the new operators to be smaller than their incumbent competitors. Smaller employment size, correspondingly, has allowed new operators to create three other organizational 'F-characteristics': flatness, fastness, and flexibility. Compared with the incumbents, the new operators have been able to avoid hierarchical, bureaucratic organizations in favour of flat organizations; they have been able to establish streamlined decision-making procedures facilitating fast decisions; and this has given them the ability to change direction or establish new directions more quickly than their incumbent competitors. Together with focus, in short, they have been able to become 'F4 Firms'.

An example is provided by Energis, the telecoms subsidiary of the National Grid, the main electricity provider in England and Wales. Table 2.4 shows the functional employment of Energis, that has become one of the main long-distance competitors to BT and more recently has moved into metropolitan area networks and has begun operations on the European continent.

At this stage, Energis, which started operations only in 1993, had sales of around £400 million, had built a network of 6,500 km covering all the principal business centres in England, Wales, and Northern Ireland, and had already joined the Financial Times stock exchange (FTSE) index 100 list of the most valuable companies.[28] Leading stock market analysts estimated that Energis

---

[28] Energis plc, *Interim Report 1999/2000*, p. 2.

Table 2.4. Employment by function in Energis

Department/function	Number	Per cent
Customer service	138	17
Executive	6	0.8
Finance, property & legal	123	15
Human resources	11	1.4
Information systems	84	11
Marketing	48	6
Network operations & engineering	256	32
Sales	129	16
Strategy and business development	3	0.4
Total	798	100

Source: HSBC, *Global Offering of 60,000,000 Ordinary Shares at a Price of £16.50 Each in Energis plc*, 22 January 1999, p. 19.

would win 11 per cent of the UK market for advanced telecoms services by 2006. In 1998, Energis already carried 60 per cent of the UK's total Internet traffic.

The Differentiation Dilemma. Although the network operators have benefited greatly from the technology supplied by the specialist technology suppliers, they have also had to confront the downside of this transaction. By depending on the specialist technology suppliers, who supply their state-of-the-art technology to anyone with the ability to pay for it, the network operators have forgone a possible source of differentiation from their competitors, namely improved products, services, and process brought about by internally produced technical change. This may be referred to as the differentiation dilemma. In short, network operators are unable to enjoy the benefits of better technologies than their rivals. All have access to essentially the same technologies.[29]

The differentiation dilemma is particularly acute for the new operators who, as we have seen, do virtually none of their own R&D.[30] The R&D-performing

[29] Of course, different network operators may choose different technologies. For example, operators competing in the local access market have the choice of optical fibre, XDSL/copper cable, coaxial cable, fixed wireless access, cellular mobile, and even satellite. However, other operators have access to exactly the same technologies.

[30] In theory, new operators do have the option of outsourcing R&D to specialist technology suppliers. In my interviews, I have come across cases where the new operator has negotiated temporary privileged access to the output resulting from outsourced R&D. After the temporary period has ended, however—usually a short period of around six months—the specialist technology supplier tends to assume full control over the technology. In practice, however, there seem to be very few examples where this has given the new operator a significant technology-based competitive advantage.

incumbents, on the other hand, have a potential advantage stemming from their in-house R&D capabilities. However, as we have also seen, the vast bulk of R&D in the New Telecoms Industry is done by the specialist technology suppliers and this accordingly limits the ability of the incumbents to achieve significant technological advantages from their in-house R&D. Nevertheless, this issue raises the important question of whether, over time, the new entrants will find that they too should be doing their own internal R&D in order to keep up in the competitive race.

The differentiation dilemma is a dilemma, precisely because in the absence of differentiation, and with substantial new entry facilitated by low entry barriers, firms are unable to earn scarcity rents. Accordingly, profit margins will be low. This raises a key question: How do network operators compete in the New Telecoms Industry? What characteristics drive competitiveness?

In some cases, a partial solution to the differentiation dilemma is available to a few network operators. For example, Qwest, as a result of Philip Anschutz's railway property rights, was able to acquire important rights-of-way that allowed the company to secure its optical fibre conduits by burying them alongside railway lines. In turn, the scarcity of these property rights allowed Qwest to earn substantial economic rents. More specifically, Qwest has been able to earn significant revenues by selling some of the capacity on its optical fibre networks to competitors Frontier, WorldCom, and GTE in the form of dark fibre.[31] Indeed, Qwest has stated that 'the sale of dark fibre [primarily to these three competitors] has financed more than two-thirds of our overall [network] construction costs'.[32] Similarly, cellular mobile and fixed wireless access companies have benefited from the natural scarcity of spectrum (to some extent paid for by the auction price of spectrum). Some competitive local exchange carriers have also argued that permission to dig up the streets and lay cables, and the sheer difficulties involved in so doing, constitute important entry barriers and, faced with competitive service tariffs, make entry relatively unattractive for late-comers.[33]

However, even after these entry barriers are taken into account, the problem for an operator attempting to differentiate its products and services remains. So, how do network operators compete in the New Telecoms Industry?

In competing with the incumbents, the new new operators enjoy a particular competitive advantage (that has to be set against the disadvantages of the new entrant *vis-à-vis* the established incumbent). This advantage stems from the so-called legacy networks of the incumbent, that is the older-generation technologies that are, inevitably, incorporated in parts of the incumbent's network. The new new operator, on the other hand, can start with a clean slate, deploying only the latest, state-of-the-art technology. Reading through the evaluations of the new new operators made by the leading stock market analysts, it is clear

---

[31] That is, unused, or unlit, optical fibre.   [32] Qwest, *Annual Report 1998*, p. 13.
[33] Author's interviews.

that considerable weight has been placed on the technological advantage of the new entrant relative to the incumbent in calculating expected future market share and earnings.[34] The problem facing the new new operator, however, is that while the latest technology may provide a competitive advantage against the incumbent, the window of opportunity thereby provided is inevitably short-lived: in due course, even newer entrants will be able to enjoy the benefits of even more recent technologies.

Furthermore, in competing with other new new entrants that have entered at the same time, an operator is unable to rely on technological superiority. In these cases, competition revolves around the attempt to provide superior quality of service—such as quicker provisioning time and quicker restoration of disrupted service—and superior customer care, such as better understanding of customer needs and a greater ability to provide solutions, on the basis of the same common technology, to customer problems. This competitive process is similar to other industries where service providers are supplied by common specialist technology suppliers, such as the airline business where airlines, supplied by the same providers of airframes, aero-engines, and in-flight entertainment systems, struggle to persuade customers that they are somehow different.

This analysis of the differentiation dilemma, however, raises the puzzle of why stock markets have valued the shares of new new entrant operators so highly. In turn, this poses the broader question of the role that financial markets have played in the New Telecoms Industry.

The Role of Financial Markets: During the Telecoms Boom[35]

In this chapter, significant weight has been put on the technological and learning regimes as an 'engine' of evolution in the Telecoms Industry. However, it is not the only engine. Another important engine of change has been financial markets.

Financial markets have influenced the evolution of the Telecoms Industries in two major ways. First, they have facilitated the entry and initial growth of new entrants, in particular, the new new entrants. Second, they have facilitated

[34] For example, in the valuations of Qwest, considerable weight has been attached to the company's competitive advantage that has followed from its more effective use than AT&T of technologies such as self-healing SONET (an optical protocol facilitating broadband capacity transmission) ring architectures, that allow traffic to be routed in two directions, thus facilitating a continuity of service in the event of a break in the optical fibre cable; advanced optical fibre and transmissions technologies that allow for OC-192 level bandwidth which operates at 10 gigabits providing the highest speeds currently available; and a 2.4 gigabit (OC-48) IP architecture that supports the most advanced data communications services.

[35] The role of financial markets during the Telecoms Boom and Bust, 1996–2002, is examined in Chapter 1.

the 're-shuffling of the capital stock'[36] that has taken place as both network operators and specialist technology suppliers with highly valued shares have used their valuable 'paper' (shares) to acquire the complementary knowledge and tangible assets of other companies. By so doing, financial markets have facilitated the process of consolidation in the Telecoms Industry that, in turn, has enabled dynamic increasing returns, economies of scale, and economies of scope to be realized.

Financial markets have facilitated the entry and initial growth of new entrants as investible resources have been made available to the owners of these companies, primarily through bond and equity markets. But, financial markets have also aided entry and initial growth through the equity valuation process. Those new entrants that have been able to convince financial analysts and investors that they have relatively attractive future prospects have been rewarded with appreciating relative share values. In turn, appreciating values have enabled these new entrants to further tap bond and equity markets on reasonable terms. Furthermore, as already noted, by making employee incentive schemes such as stock options lucrative, financial markets have reinforced the operation of labour markets, enabling new entrants to attract necessary human resources.

Acquisitions have also been facilitated by the stock market valuation process.[37] For example, WorldCom's acquisition of MCI, Vodafone's acquisition of Airtouch and Mannesmann, Qwest's acquisition of US West, and Global Crossing's acquisition of Frontier were paid for largely with shares of the acquiring company. In this way, these companies used their shares, highly valued by stock markets, as a 'currency' with which to finance their acquisitions.

In Table 2.5, the market capitalization of the main US new new entrants is examined and compared to the incumbents and some of the major specialist technology suppliers.

Table 2.5 shows the remarkable market capitalization of the US new new entrants, shown in bold, who were only established in the late 1980s and early 1990s. These figures are for mid-1999 and at this time the new new entrants were together worth approximately the same as BT.[38]

---

[36] This is a concept that comes from the Austrian economist, Luwig Lachmann.

[37] The high-yield bond market has also played an important role in providing the finance for expanding telecoms operating companies to extend their networks. One advantage of resorting to bonds rather than equity in funding expansion is that the dilution of ownership and control can be limited. The bonds, however, are high yielding as a result of the high risk attached to loans to new entrant telecoms operators.

[38] These figures are changing almost weekly. For example, they do not reflect Vodafone's acquisition of Airtouch and Mannesmann, Qwest's acquisition of US West, Global Crossing's acquisition of Frontier. In March 2000, the market capitalization of Cisco briefly exceeded that of Microsoft, hitherto the world's most valuable company. The figures in Table 2.5, however, suffice to make the main point emphasized here, namely the stock market valuation process that has significantly increased the value of the shares of the new new entrant fixed network operators.

Table 2.5. Market capitalization of some new entrants and other selected companies, 1999

Companies	Rank	Market value ($ bn)	Country
Microsoft	1	407	US
AT&T	7	186	US
Cisco	9	174	US
NTT	13	157	Japan
MCI–WorldCom	**14**	**152**	**US**
Lucent	16	150	US
Deutsche Telekom	23	115	Germany
BT	26	107	Britain
NTT Docomo	27	106	Japan
SBC Communications	31	100	US
France Telecom	43	80	France
Telecom Italia	58	67	Italy
Nortel Networks	84	50	Canada
Qwest	**146**	**30**	**US**
Level 3	**172**	**27**	**US**
Williams	**204**	**22**	**US**
Global Crossing	**244**	**20**	**US**
NTT Data	255	19	Japan
Global Telesystems	**716**	**6**	**US**

Source: *Business Week*, 12 July 1999.

It is clear, therefore, that the process of stock market valuation of companies has played a key role in the Telecoms Industry, justifying the inclusion of financial markets as one of the 'engines' driving the evolution of the industry. However, since stock market valuation has played such an important role it is necessary to inquire further into how these 'values' are determined.[39]

Inventing the Value of Telecoms Company Shares.[40] In principle, the value of a company's shares is determined by the present discounted value of that company's future earnings. Accordingly, the would-be valuer of a company's shares must decide on what discount rate to use and on how to determine the company's future earnings. Ignoring the discount rate, focus will be on future earnings and, in particular, on the future earnings of the new new entrant network operators.[41]

---

[39] The theory of value has played a key role in economic thought from the Greeks through the classical political economists such as Adam Smith, David Ricardo, and Karl Marx to the neoclassical economists from the late-nineteenth century. It may well be argued that the 'valuation' of telecoms and Internet shares at the turn of the twenty-first century adds further relevant material for the debate about the determinants of 'value'.

[40] The discussion in this section is closely related to that of the Telecoms Boom and Bust in Chapter 1.

[41] The same issues regarding valuation, however, also apply to new Internet companies and other companies that have little relevant track record.

The problem that immediately arises in attempting to calculate the future earnings of the new new network operators is one of *uncertainty*.[42] For example, uncertainty arises regarding the ability of the new company to organize and manage its entry and initial expansion; regarding future market demand and the proportion of the market that the new company will be able to address and win; regarding the number and strength of future competitors; regarding the extent of the threat posed by alternative technologies; regarding the importance of future substitutable services; etc. Furthermore, to make matters worse, the new new entrants, by definition, begin without any track record on which analysts and investors may base judgements and, in addition, typically make significant losses in their set-up period as a result of the substantial fixed costs involved in constructing their networks, costs that are incurred in advance of compensating revenue being generated.

Grappling with the problem of valuing the shares of a new new operator, the financial analyst might well be forgiven for a feeling of bewilderment at the concept of the 'efficient markets hypothesis'. This hypothesis holds that a capital market is efficient if 'it fully and correctly reflects all relevant information in determining security prices. Formally, the market is said to be efficient with respect to some information set, x, if security prices would be unaffected by revealing that information'.[43]

The first problem the financial analyst may have with this concept is that 'information', by definition, refers to the past; it is not possible to have information about the future. Yet, the problem arising in valuing a new new operator's shares relates essentially to future magnitudes. Second, the current information set yields contradictory inferences regarding many of the key variables with which the financial analyst is concerned. For example, Chris Gent, CEO of the largest global mobile company, Vodafone, is adamant in his belief that mobile communications will seize a significant part of the market from fixed communications. Bernard Ebbers, CEO of WorldCom, however, was equally vehement in his rejection of this belief.[44]

In short, decision-makers in the Telecoms Industry usually confront what has been termed 'interpretive ambiguity' in attempting to calculate the implications of the current information set. Interpretive ambiguity may be defined as existing when the information set is capable of yielding contradictory

---

[42] Many years ago, Frank Knight (1921) drew a crucial distinction between risk and uncertainty. In the case of risk, probability distributions can be calculated on the basis of present data that can reasonably be expected to be valid for the future. These probability distributions, for example, provide the basis for the operations of the insurance industry. In the case of uncertainty, however, such probability distributions cannot be derived. This distinction, unfortunately, is often neglected. As Knight (1921) put it, 'a *measurable* uncertainty, or "risk" proper... is so far different from an *unmeasurable* one that it is not in effect an uncertainty at all. We shall accordingly restrict the term "uncertainty" to cases of the non-quantitative type' (p. 20).

[43] Malkiel (1987: 120).

[44] Although there was some evidence of a shift in belief when Ebbers attempted (unsuccessfully) to acquire Sprint, largely as a result of the attraction of that company's mobile network.

inferences regarding what will happen.[45] Under such circumstances, the decision-maker simply has no alternative but to construct his or her own 'visions' of what will happen in the future, based on personal beliefs and expectations. Rather than being able to bask in the sunshine provided by the notion that the markets (of which she or he is an organic part) are 'fully and correctly reflecting all information', however, the financial analyst will also be uncomfortably aware of the possibility of 'vision failure', that is of being wrong.

So, how does the financial analyst proceed in the light of this incomplete information and interpretive ambiguity?[46] The answer, as a reading of the company reports of the leading telecoms financial analysts readily shows, is that the analyst 'invents' the key assumptions that drive the calculations of future cash flows and earnings. There is little pretence in these reports that the 'inventions' made 'fully and correctly reflect all information'. Rather, the language of the analyst usually disarmingly betrays the discomfort that all decision-makers feel under conditions of interpretive ambiguity.[47]

But, it must be said that the strength of these whole valuation processes lies, not in its objectivity, but in the fact that the 'visions' that guide this invention process, and the beliefs and expectations that underlie these visions, are made explicit and, therefore, are subject to the possibility of disagreement on the part of those with different visions. In this way, the conditions are created for

---

[45] See Fransman (1999). As Knight (1921) noted, 'we do not react to the past stimulus, but to the "image" of a future state of affairs.... We *perceive* the world before we react to it, and we react, not to what we perceive, but always to what we *infer*' (p. 201).

[46] Note that the problem here is not Herbert Simon's problem of 'bounded rationality' resulting from excessive information, which is greater than the ability of the human mind to process that information. Simon (1957) defines bounded rationality in the following way: 'the capacity of the human mind for formulating and solving complex problems is very small compared with the size of the problems whose solution is required for objectively rational behavior in the real world—or even for a reasonable approximation to such objective rationality' (p. 198). The problem being dealt with here is not that the 'capacity' of the human mind is 'very small' relative to the large amount of information that must be processed. Rather, the problem stems from the fact that the information set yields interpretive ambiguity in the sense that it yields contradictory inferences. Under these circumstances, different 'rational' people may well arrive at contradictory conclusions regarding what to infer and, therefore, how to act.

[47] The following quotations are typical. They come from financial analysts in the research department of one of the best-known financial companies in a valuation of one of Europe's best-performing new new entrants:

We are using a five-year DCF [discounted cash flow] model...in order to value [X]. Quite obviously, given [X's] relative immaturity, it is impossible to value the company using conventional earnings ratios such as price/earnings, price/earnings relative, EPS [earnings per share] growth and even firm value (market capitalization plus net debt)/EBITDA [earnings before interest, tax, depreciation, and amortization].

We are using a discount rate of 14%.... At present we have assumed that further finance is provided through debt. It is entirely possible that the company could choose to issue equity, but we have more confidence in our ability to predict future interest rates than to predict at what price future equity could be sold. This does, of course, mean that interest expense could be overstated.

'a thousand flowers blooming, a thousand thoughts contending'—all in all a reasonable way of dealing with conditions of interpretive ambiguity.

Stock market valuation, however, involves more than the attempted prediction of future earnings. It also involves the attempted calculation of how other investors will react to the same ambiguous set of information. In effect, therefore, there are two evaluation questions that are being tackled. First, what do we, the decision-makers, expect the future earnings of the company would be? Second, what do we think other decision-makers (e.g. financial analysts and investors) will say and do regarding the company?

Given interpretive ambiguity/incomplete information, these two questions are independent of each other. For example, I may believe that a company does not have good relative prospects but still buy its shares because I believe that sufficient others will disagree and will buy the shares in the future. In this way, 'circular expectations', that is, expectations of other peoples' expectations, may enter into the valuation process even when the link between the expectations and the supporting information is tenuous. Under these conditions 'fashion', rather than 'information', may drive share prices. It is this kind of valuation process that has driven the market value of many Internet start-ups around the world. However, tenuous as these stock market values may at times be, they have had a significant impact on 'real' variables and, therefore, on the evolution of the Telecoms Industry.

An important further factor that has influenced the rapid appreciation in the share values of many telecoms companies has been the 'explosion in demand for data communications' as a result of the global adoption of the Internet. However, here too the invention of beliefs and visions has played a role. This is most apparent in the paucity of accounts that couple the explosion in demand with the accompanying explosion in supply and analyse the combined effects. A further key variable is the price elasticity of demand, since a fall in the price of data-carrying capacity as a result of the combined effects of the demand and supply of capacity may nonetheless, if the price elasticity is high enough, be accompanied by an increase in the total revenue of all the sellers of capacity. An additional issue, crucial in the valuation of individual company shares, is how successful a company will be in winning its share of the increased total market

---

We are using a terminal EBITDA multiple of 10 [in order to calculate the final share price as a multiple of the EBITDA]. Such a multiple suggests substantial growth in EBITDA and EPS even after 2003. We believe (to the best of our ability to predict what will happen in 2004 and beyond) that [X] will be increasing EBITDA at 12%–15% and EPS at 15%–20% post-2003. We believe that 10× terminal multiple is reasonable relative to growth profiles, the size of [X's] opportunity, its network/technology advantages (fibre, SDH) and [X's] EPS/EBITDA growth post-2003.

Nonetheless, we realize the inherent volatility and uncertainty in attempting such a valuation and we have tried to perform several cross-checks in order to validate our assumptions and methodology, including sensitivity analysis of multiple and discount rates.

revenue. These factors, however, are seldom introduced explicitly into the valuations.[48]

THE INTERNET AS A NEW PARADIGM AND THE BIRTH OF THE INFOCOMMUNICATIONS INDUSTRY

In the early 1990s, the Internet emerged as a commercial force, creating an alternative way of delivering the same or similar services to those provided over the conventional circuit-switched telecoms networks and, in addition, a host of new services. But the Internet was far more than just an alternative platform; it was nothing less than a radically new paradigm in the area of both information and communications, changing fundamentally the way in which people would think of problems and solutions in this field. Furthermore, by inserting itself into the very fabric of the Telecoms Industry, the Internet brought about the metamorphosis of this industry into what will be termed here the Info-communications Industry. In this section, the emergence of the Internet and its impact on the Telecoms Industry will be examined.

The Emergence of the Internet[49]

In Table 2.6, the main landmarks of the Internet are shown.

ARPA and Time-shared Computing. The institution that gave birth to the net-work that was eventually to evolve into the Internet was the Advanced Research Projects Agency (ARPA) that was established in 1958 in the US in response to the Russian launch of Sputnik. From 1962 to 1964, ARPA, under J. C. R. Licklider, encouraged through its funding the development of time-sharing computer systems at leading universities and government-funded research laboratories in the US. 'One of Licklider's strong interests was to link these time-shared computers together through a widespread computer network.'[50] This interest was motivated by the high cost of time-share computers and the desire to make more efficient use of this resource. Licklider's interest and the considerable research funding controlled by ARPA and its Information Processing Techniques Office (IPTO), established in 1962 to further advanced research in computing, served to stimulate discussion and debate around the questions of how to

---

[48] The points made in this paragraph are related to the causes of the Telecoms Boom and Bust, 1996–2002, in Chapter 1.

[49] This section draws heavily on the account of the evolution of packet switching given by Lawrence Roberts, who joined ARPA in January 1967 and managed its computer research programmes—see Roberts (1978)—and on Janet Abbate's (1999) excellent *Inventing the Internet.* See also Norberg and O'Neill (1996).   [50] Roberts (1978: 1308).

Table 2.6. Landmarks in the evolution of the Internet

Date	Event
1950s	1958—ARPA founded in response to Sputnik.
1960s	Early 1960s—Packet switching invented independently by Paul Baran, Rand Corporation in the US, and Donald Davies, National Physical Laboratory, UK, based on the notion of 'message switching' that went back to the postal and telegraph systems.
	1962—ARPA's IPTO founded.
	1967—ARPANET project started.
1970s	1972—ARPANET demonstrated at the First ICCC.
	First commercial packet switching network introduced.
	E-mail starts to be widely adopted.
	1973—Robert Kahn approaches Vinton Cerf to develop a system for Internetworking and they outline the basic Internet architecture.
	1974—Initial version of TCP specified.
	AT&T declines to take over operation of ARPANET.
	1975—Ethernet created by Robert Metcalfe.
	1977—ARPANET demonstrates its first multinetwork connection, connecting the networks ARPANET, PRNET, and SATNET.
	1978—IP established.
1990s	1990—First incarnation of the World Wide Web, created by Tim Berners Lee, Robert Cailliau, and others at CERN, Switzerland.
	1993—Marc Andreesen and team develop improved Web browser, Mosaic.
	1994—Andreesen and team set up commercial version of Mosaic, Netscape.
	1995—On 26 May 1995, Bill Gates issues memo, 'The Internet Tidal Wave', that acknowledges that Microsoft will have to adapt all its systems to the Internet.

construct computer networks to facilitate interactive computing between time-shared computers.

Clearly, if time-shared computers were to be linked, a communications system would be needed in order to facilitate the flows of data between the mainframe computers.[51] This, in turn, raised the question of what kind of communications system would be suitable.

Circuit switching versus Packet switching. The obvious way to link distant time-shared computers was through leased telephone lines. However, the cost of using these leased lines was high. Apart from AT&T's monopoly over leased lines, a major determinant of the high cost was the technology that was used, namely a *pre-allocation* technique called *circuit switching* according to which a fixed bandwidth is pre-allocated for the duration of the connection. While this

---

[51] As Roberts put it, 'the interest in creating a new communications system grew out of the development of time-sharing and Licklider's special interest in the 1964–65 period' (p. 1308).

technique was fairly well suited for the transmission of voice calls, it was not particularly well adapted for the transmission of interactive data traffic that occurs in short bursts and that can result in as much as 90 per cent of the pre-allocated bandwidth being wasted. This, in turn, raised the question of whether there were any more cost-effective alternatives.

It had long been realized that an alternative to pre-allocation existed in the *dynamic allocation* of transmission bandwidth. This method had traditionally been used in postal mail systems and the telegraph. The method involved collecting and storing messages at a node in the network and then, when sufficient messages had been accumulated, sending the messages in bulk on to the next node in the network. This made better use of transmission capacity compared to the pre-allocation technique that was wasteful since it allocated capacity even when no messages were being transmitted. The problem with dynamic allocation, however, was that many sorting and routing decisions had to be made by human operators and this entailed a high cost in using this technique.

The advent of relatively inexpensive computers, however, created the possibility of removing this obstacle to dynamic allocation. It is here that the computer revolution, and the linked revolutions in computer storage and processing, enters the story of the evolution of the Internet. In 1965, the time when discussions were being held regarding an appropriate communications system for linking time-shared computers, DEC introduced its PDP-8 minicomputer that would drastically reduce computing costs compared to the mainframes that then dominated. This computer revolution gave a new breath of life to dynamic allocation. In the words of Roberts: the 'economic tradeoff [between pre-allocation/circuit switching and dynamic allocation/packet switching] is simple: if lines are cheap, use circuit switching; if computing is cheap, use packet switching' (p. 1307). However, Roberts also noted that this trade-off, recognizable with hindsight, was by no means acknowledged at the time: 'Although today this seems obvious, before packet switching had been demonstrated technically and proven economical, the tradeoff was never recognized, let alone analyzed' (p. 1307).

The alternative of dynamic allocation/packet switching as an appropriate technology for the communications system linking time-shared computers, however, did not emerge only as a theoretical possibility, based on old postal and telegraph systems. Coincidentally, research in this area had been undertaken independently by two researchers, one in the US and one in the UK. It was their research, and the feasibilities that it demonstrated, rather than the purely theoretical superiority of dynamic allocation, that paved the way for the adoption of packet switching as the basis for the communications system chosen for ARPANET, the network that would be constructed to link the time-shared computers in ARPA's computer research programmes.

At the Rand Corporation, Paul Baran in the early 1960s did research for the US Air Force on a military communications system for data and voice. 'The Air

Force's primary goal was to produce a totally survivable system that contained no critical central components.'[52] The injunction against centralized control of the system followed from the desire to make the system less vulnerable (i.e. 'survivable') to Russian attack. Baran's research, published in August 1964, proposed a fully decentralized (i.e. distributed) communications system based on packet switching/dynamic allocation. However, his 'report sat largely ignored for many years until packet switching was re-discovered and applied by others'.[53]

In the UK in the early 1960s, at the government-funded National Physical Laboratory, Donald Davies was also developing a communications system based on packet switching/dynamic allocation designed for interactive computing.[54] In autumn 1965, Davies sponsored a conference reporting on his results that was attended by Lawrence Roberts (who would join ARPA in January 1967 and assume management of its computer research programmes) and others from MIT. It was at this conference that Roberts and his colleagues decided that dynamic allocation should be used in ARPANET. 'Almost immediately after the 1965 meeting, Donald Davies conceived of the details of a store-and-forward packet switching system, and in a June 1966 description of his proposal coined the term "packet" to describe the 128-byte blocks being moved around inside the network. Davies circulated his proposed network design throughout the UK in late 1965 and 1966. It was only after this distribution that he discovered Paul Baran's 1964 report.'[55]

Communications Engineers versus Computer Professionals. What was obvious to some of the key computer scientists in the National Physical Laboratory and the ARPA computer research programme, however, was by no means obvious to those in the Telecoms Industry whose 'visions', and the beliefs and interpretations they embodied, had little room for a reversion to what they disparagingly saw as outdated dynamic allocation techniques. In the words of Lawrence Roberts:

In the early 1960s, preallocation [i.e. circuit-switching] was so clearly the proven and accepted technique that no communications engineer ever seriously considered reverting to what was considered an obsolete technique, dynamic allocation. Such techniques had been proven both uneconomic and unresponsive 20–80 years previously, so why reconsider them? The very fact that no great technological breakthrough was required to implement packet switching was another factor weighing against its acceptance by the engineering community.

It was only from a group of outsiders, with a fundamentally different starting point, set of problems, and set of beliefs and interpretations, that a new alternative technology, based on dynamic allocation, could emerge: 'What was required was the total re-evaluation of

---

[52] Roberts (1978: 1308).   [53] *Ibid.*   [54] Abbate (1999: 23).
[55] Roberts (1978: 1308). According to Abbate (1999), it was in March 1966 that Davies first presented his ideal publicly to an audience of computing, telecoms, and military people and a man from the British Ministry of Defence gave his surprising news about Baran's 1964 report (p. 27).

the performance and economics of dynamic-allocation systems, and their application to an entirely different task. Thus, it remained for outsiders to the communications industry, computer professionals, to develop packet switching in response to a problem for which they needed a better answer: communicating data to and from computers'.[56]

In evaluating Roberts's evaluation of the respective role played by 'computer professionals' and 'communications engineers' it is worth being reminded of the research agenda that existed at the time in the field of telecoms switching. In the mid-1950s, the first electronic space-division switching research programme incorporating stored program control (SPC) was introduced by Bell Laboratories. This resulted in the first trial electronic SPC switch in 1960 and the first commercial switch, AT&T's No. 1 ESS in 1965.

But no sooner were the first electronic space-division switches introduced than they were in the process of being displaced by the next generation of central office switches, namely by time-division digital switches. Although the first research on time-division switching went back to the work of Reeves in 1938 and Deloraine in 1945, it was only in 1959 that the Essex Project in Bell Labs demonstrated the feasibility of time-division switching. This was followed by the development of pulse-code modulation (PCM) transmission systems in the early 1960s and the beginning of research in Bell Labs on AT&T's first digital switch, the No. 4 ESS, the first laboratory model of which was introduced in November 1972. In 1976, Northern Telecom (now Nortel) began development of its first relatively small digital switch, the DMS, which was commissioned in 1977, about four years earlier than AT&T's equivalent switch, the No. 5 ESS, that was introduced in 1982. As this brief account makes clear, therefore, the telecoms engineering community were very much focused on the development of electronic space-division switching, and then time-division digital switching, both within the broader context of circuit switching, at the time that the debates were occurring in the ARPA community about the feasibility of packet switching for the communications system that would connect time-shared computers.[57]

In October 1972, the first packet-switched network was publicly demonstrated at the first meeting of the International Conference on Computer Communications (ICCC) in Washington, DC when a complete ARPANET node was installed in the conference hotel with about 40 active terminals permitting access to dozens of computers in various parts of the US. This provided proof that packet switching could really work and managed to convince many working in the networking field. However, some still remained to be persuaded. According to Roberts, in an article published in 1978, 'AT&T evidenced even less interest in packet switching than many of the PTT's in other countries. AT&T and its research organization, Bell Laboratories, have never to my knowledge published

---

[56] Roberts (1978: 1307).

[57] This discussion on telecoms switching comes from Fransman (1995), ch. 3, specifically from fig. 3.1, p. 47, and fig. 3.3, p. 50. See also Chapuis and Joel (1990).

any research on packet switching. ARPA approached AT&T in the early 1970s to see if AT&T would be interested in taking over the ARPANET and offering a public packet switched services, but AT&T declined'.[58]

The Internet Protocol. It was in this way that a new communications paradigm was born based on packet switching and offering a radically new approach to the communication of both data and voice.[59] In 1968, virtually all interactive data communication networks were circuit switched. By 1978, however, virtually all new data networks being built throughout the world were based on packet switching, a remarkable rate of diffusion for a radically new technology.[60]

In the spring of 1973, ARPA researchers Robert Kahn and Vinton Cerf got together to consider how to interconnect *dissimilar* networks.[61] This led eventually to Transmission Control Protocol (TCP)/IP. As shown in Table 2.6, in 1974 the initial version of TCP was specified. In 1978, 'They proposed splitting the TCP protocol into two separate parts: a host-to-host protocol (TCP) and an internetwork protocol (IP).... IP would simply pass individual packets between machines (from host to packet switch, or between packet switches); TCP would be responsible for ordering these packets into reliable connections between hosts'.[62] In this way, the transfer of packets across different networks, using different technologies, was facilitated.

The Proliferation of the Internet. When did the realization first dawn that the Internet has become a Schumpeterian tidal wave of creation–destruction? One way to tackle this question is to examine the reactions of Vinton Cerf, often referred to as one of the fathers of the Internet, who, as noted earlier, co-wrote the IP with Robert Kahn and made major contributions to the TCP. Cerf has said that in 1977, he 'Assumed that [ARPANET] was a research project that would probably never get bigger than 128 networks'.[63] By 1996, the Internet was a network of some 50,000 networks.

When did Cerf first realize that the innovations that he and his co-workers made would form the basis for a fundamental shift in the way in which we think

---

[58] Roberts (1978: 1310). See Abbate (1999: 137) for further discussion.

[59] Roberts's prediction in 1978 regarding voice communications by packet switching are apropos: 'The economic advantage of dynamic-allocation over pre-allocation will soon become so fundamental and clear in all areas of communications, including voice, that it is not hard to project [predict?] the same radical transition of technology will occur in voice communications as has occurred in data communications... the obvious solution would be an integrated packet switching network that provides both voice and data services....Given the huge fixed investment in voice equipment in place today, the transition to voice [packet] switching may be considerably slower and more difficult. There is no way, however, to stop it from happening' (p. 1312). In 1995, an Israeli-owned startup, based in Silicon Valley, Vocaltech, introduced the first software allowing for voice over the Internet. [60] Roberts (1978: 1307).

[61] See Abbate (1999: ch. 4), From ARPANET to Internet. [62] Abbate (1999: 130).

[63] This and the following quotations from Vinton Cerf come from Cerf (1996).

Table 2.7. The Infocommunications Industry: a layer model[64]

Layer	Activity	Example companies
VI	*Customers*	—
V	*Applications Layer, including contents packaging* (e.g. web design, on-line information services, broadcasting services, etc.)	Bloombergs, Reuters, AOL-Time Warner, MSN, Newscorp, etc.
IV	*Navigation & Middleware Layer* (e.g. browsers, portals, search engines, directory assistance, security, electronic payment, etc.)	Yahoo, Netscape, etc.
III	*Connectivity Layer* (e.g. Internet access, Web hosting)	IAPs and ISPs
IP interface		
II	*Network Layer* (e.g. optical fibre network, DSL local network, radio access network, Ethernet, frame relay, ISDN, ATM, etc.)	AT&T, BT, NTT, WorldCom, Qwest, Colt, Energis, etc.
I	*Equipment & Software Layer* (e.g. switches, transmission equipment, routers, servers, CPE, billing software, etc.)	Nortel, Lucent, Cisco, Nokia, etc.

of information and communications? The answer is between 1989 and 1991 as Cerf makes clear:

We now know that the lift-off point for exponential growth [in the Internet] came around 1988, though it wasn't obvious at the time. I certainly didn't sense it in the early days of research, 22 years ago. But I remember vividly walking into a trade show in 1989, one of the InerOp shows in San Jose, and looking at the booths, and this was the first time that I had seen really expensive booths being put up. It was obvious that the vendors were spending very significant dollars and that there was something serious going on. And then the second epiphany, if you can call it that, came a couple of years later when the first commercial service companies started to show up. This really confirmed my increasing expectation that this was turning into a serious business.

And how did he feel when he realized what was happening? 'My first reaction was something like, "Holy Shit"'.

In 1990, as shown in Table 2.6, the first incarnation of the World Wide Web emerged, created by Tim Berners Lee, Robert Cailliau, and others at CERN, the particle-physics research establishment in Switzerland. In January 1993, Mosaic was introduced, the first World Wide Web browser, based on research done at the University of Illinois. In April 1994, Mosaic Communications was established, a firm that soon became Netscape Communications Corp. that was

---

[64] This diagram of the Layer Model is also used in Chapter 1 in order to analyse the Telecoms Boom and Bust, 1996–2002. I would like to express my gratitude to Sadahiko Kano for his contribution in developing this version of the layer model. For further discussion of this layer model, see http://www.TelecomVisions.com.

floated on the stock exchange on 8 August 1995 (the shares increasing in value on the first day from $28 to $58).

There is evidence that Bill Gates began to understand the implications of the Internet only in 1995. Although Gates and Microsoft were immersed in the development and shipping date of Windows 95 (eventually released on 24 August 1995) and until mid-1995 did not pay too much attention to the Internet, Gates' vision of the future of computing and Microsoft had come to encompass the importance of networked computing. On 6 October 1994, Gates wrote a memorandum titled 'Sea Change' that spelled out plans for networked computing Microsoft. Earlier, in May 1993, Gates had approved work on Marvel, an online service that would be offered by Microsoft (and later became MSN). Marvel, however, was not intended to be Internet-compatible.

Until early to mid-1995, however, it is clear that Gates saw Microsoft as the dog of networked computing and the Internet, at best, as a rather insignificant tail. In Gates' own words: 'I wouldn't say it was clear [at this time] that [the Internet] was going to explode over the next couple of years. If you'd asked me then if most TV ads will have URLs [Web addresses] in them, I would have laughed'.[65]

By the autumn of 1995, however, some 20 million people were accessing the Internet without using Microsoft's software and on 23 May 1995, Sun's Java software language was officially released which, being platform-independent, further threatened to bypass Microsoft's systems. By the latter date, however, the alarm bells that had for some time been ringing inside Microsoft (in the form of several younger staff, more in touch with the rapidly growing adoption of Internet technologies, particularly on American campuses) were heard by Gates and the rest of the leadership. On 26 May 1995, Gates issued a memorandum, 'The Internet Tidal Wave', that finally confirmed his conversion to the view that the Internet had become the dog, with Microsoft, after all, only its tail.

The Birth of the Infocommunications Industry

How has the Internet changed the Telecoms Industry and transformed it into the Infocommunications Industry? The Internet has had four major effects on the Telecoms Industry that together have fundamentally changed this industry. Table 2.7 shows in the form of a layer model the main features of the Infocommunications Industry (which may be contrasted with the features of the Old Telecoms Industry, shown in Table 2.1).

First, the Internet has established that packet switching and the IP networks in which it is embodied constitute a superior technology compared with circuit-switched technologies, not only for data but also, as Lawrence Roberts correctly foresaw at least by 1978 (see above), for voice. Indeed, to go even further, the

---

[65] *Business Week*, 15 July 1996, pp. 38–44.

Internet has created a fundamentally new paradigm for the understanding of information and communications problems and solutions.

Second, as shown in Table 2.7, TCP/IP has created a bridge, facilitating easy and cheap interoperability across radically different networks using radically different technologies. Kavassalis *et al.* (1998) have explained the importance of the TCP/IP interface using the analogy of containerization in the Transport Industry. Pre-containerization, high costs were involved in moving goods between different transport networks such as road, rail, ship, and air. The advent of containerization, however, facilitated a smooth interface between these networks, increasing the degree of interoperability between them and significantly lowering the cost of interoperability. Likewise, TCP/IP has facilitated the transfer of bits across the different networks, embodying significantly different technologies, of the Network Layer (Layer II).

TCP/IP has produced several important consequences:

1. Easy and cheap global communications across a huge number of interconnected networks.
2. Greater effect to global standardization based around the IPs and practices.
3. A global knowledge-base, facilitating the creation of further knowledge (as will be discussed in more detail below).
4. Competition between networks, technologies, and services has been greatly increased.

Third, TCP/IP has provided a platform for three largely new layers of services, as shown in Table 2.7. Not only this, it has created the possibility for the emergence of a new industrial category of *facilities-less service providers*, specialized in one or more of the service layers, who are able to provide services using this platform while ignoring what else goes on in the network layer, Layer 2. In turn, this has created new potential for the Infocommunications Industry to become *vertically-specialized* like the computer industry did from the 1980s.[66] For example, in Layer 3, the function of connectivity is provided and with it new services, such as Internet access and Web hosting. In Layer 4, the function of navigation is provided and navigational systems such as browsers, portals, and search engines. Layer 4 also contains the 'middleware' that, sitting on top of Layers 1–3, provides the software that facilitates the applications in Layer 5, for example, software systems that provide security (such as fire walls) and facilitate electronic payment.

The fourth way in which the Internet has changed the Telecoms Industry is through the integration of the Computer Industry that it has facilitated, hence justifying the nomenclature, the Infocommunications Industry. As shown in Table 2.7, computer hardware and software and computer networks fit into the

---

[66] Whether the Infocommunications Industry will become more vertically specialized, to what extent and when, and what the implications are for incumbents, new operators, and specialist facilities-less service providers, are key questions, the causes and implications of which are addressed in http://www.TelecomVisions.Com.

Infocommunications Industry in all the layers. For example, Layer 1 contains computers such as routers and servers and software systems such as billing software. The Internet, as a 'network of networks', is an integral part of the Network Layer, Layer 2. All the services, provided in the service layers, Layers 3–5, depend on computer hardware and software. At the same time, elements of both the Old and the New Telecoms Industry have also been integrated into the layers of the Infocommunications Industry, particularly in Layers 1 and 2. From the perspective of the layer model of the Infocommunications Industry, therefore, it is possible to give a specific meaning to the widely used (perhaps overused) concept of 'convergence' between computing and telecommunications.

To conclude, the Internet that emerged from the attempts to link time-shared computers, an event that initially was remote to the Telecoms Industry, has fundamentally transformed this industry, turning it into the Infocommunications Industry.

The Technological and Learning Regimes in the Infocommunications Industry

The innovation system in the Infocommunications Industry has also undergone a fundamental transformation. Some of the important changes are shown in Table 2.8 that contrasts the innovation systems in the Old Telecoms Industry, shown earlier in Table 2.2, and the Infocommunications Industry.

As is apparent from Table 2.8, the innovation system in the Infocommunications Industry differs fundamentally from that which existed in the Old Telecoms Industry. To begin with, in the Infocommunications Industry, the innovation system is open in the sense that virtually anyone can create innovations within the industry. In marked contrast, in the Old Telecoms Industry, the innovation process was open only to the monopoly network operator and its favoured suppliers. The barriers to entry into the innovation system (i.e. the barriers constraining individuals or firms from becoming innovators) in the

Table 2.8. The innovation systems in the Infocommunications Industry and the Old Telecoms Industry

Infocommunications Industry	Old Telecoms Industry
Open innovation system	Closed innovation system
Low entry barriers	High entry barriers
Many innovators	Few innovators
Common knowledge base	Fragmented knowledge base
High-powered incentives	Medium-powered incentives
Rapid, concurrent, innovation: New forms of innovation (e.g. concurrent cooperative innovation by remote innovators)	Slow, sequential, innovation: Research–prototype–trials–cutover

Infocommunications Industry are low. Entry is greatly facilitated by the fact that there is widespread common knowledge of the main operating systems, software languages, and protocols that are used in the various layers of the industry. This common knowledge is largely the result of globalized de facto standardization (e.g. TCP/IP, hypertext markup language (html), or wireless application protocol (WAP)). In the Old Telecoms Industry, as noted earlier, many standards and practices differed markedly from country to country, resulting in a fragmented knowledge base. The increasing importance of software, together with the common knowledge base, and the relatively low cost of producing many software applications has meant that there are a large number of software innovators in the Infocommunications Industry.[67]

Furthermore, in this industry there are high-powered incentives to innovate. Internet-related innovations have a particularly large potential global market and successful innovations may be extremely richly rewarded. With so many innovators competing with each other, the rate of innovation is much faster than it was in the Old Telecoms Industry. In addition, the very nature of packet-switched networks has facilitated concurrent, rather than sequential, innovation. In the Old Telecoms Industry, a laborious process of trials was necessary *before* new equipment could be introduced into the network. If equipment in a circuit-switched network fails, the service on that network is disrupted. It was for this reason that exceptionally high reliability was a necessary requirement in the Old Telecoms Industry and, accordingly, equipment had to be exhaustively tested. However, in a packet-switched network the fact that packets can be routed in many alternative ways adds greater robustness to the network and means that many kinds of equipment and software can be tested on-line at a far earlier stage in the development process than was possible in the Old Telecoms Industry. This too has speeded up the innovation process.

In the Infocommunications Industry, new forms of innovation have been created using the Internet as a ubiquitous platform for innovation. One significant example is the concurrent cooperative development of the UNIX-based operating system, Linux, by a large number of remotely located co-innovators who do not even know each other and have no visible hand of coordination. The main features of this remarkable new process of innovation are summarized in Table 2.9.

MOBILE COMMUNICATIONS

In the late-1990s, civilian cellular mobile communications took off around the world, becoming the most rapidly growing telecommunications service. The UK company, Vodafone, acquired Airtouch of the US in 1999 and Mannesmann

---

[67] For the example of innovators in the field of live video over the Internet, see Fransman (2000*a*).

Table 2.9. Cooperative innovation by remotely located innovators: the case of Linux

Product	UNIX-based operating system
Founder	Linus Torvalds, Finnish, began as university project in 1991
Competitors	Microsoft's Windows NT
Price	Linux: Free on Internet; $50 on CD
	Windows NT: $800
Distribution	Given away on Internet, i.e. a Public Good, therefore cannot be bought by Microsoft. Many users of Linux make their own additions to the operating system which they also put, free, into the public domain.
Format	Linux: Source code—open, expandable by user
	Windows NT: Binary code—unintelligible to user
Applications	Mainly servers
Customers	Linux: 7 million, including Netscape, Intel, IBM
	Microsoft: 300 million
Performance	Linux in 1998, the fourth most often installed version of UNIX

of Germany in 2000, becoming the largest global mobile operator. NTT's majority-owned mobile subsidiary, NTT DoCoMo, introduced the first mobile Internet access service, i-mode, in February 1999. By March 2000, the company had attracted 5 million customers in Japan. Shortly thereafter, NTT DoCoMo became the most valuable company on the Tokyo Stock Exchange and one of the most valuable telecoms companies in the world, even more valuable than its parent, NTT. In the late-1990s, efforts began to develop a global standard for so-called Third Generation mobile, capable of providing Internet access at speeds of 2 megabit per second compared to the 9.6 kilobit offered by Second Generation systems.

Interpretive Ambiguity and Mobile Communications

At the turn of the millennium, it seemed obvious that consumers in many circumstances had a strong preference for mobile communications. Predictions of future penetration rates—that reached as much as 50 per cent in the most advanced countries—became commonplace. A decade and a half earlier, however, considerable interpretive ambiguity reigned regarding consumer preferences in this area, as is clear from Table 2.10.

Co-evolving Consumer Demand

As Table 2.10 makes clear, it is inappropriate to see consumer tastes and preferences as always fully formed, with firms responding to the consumer

Table 2.10. Interpretive ambiguity and mobile communications

1970s–1980s	Despite inventing the cellular mobile system, Bell Labs downgrades research on radio communications that it deems to be an inferior transmission technology.
1984	At the time of its divestiture, a Senior Executive of AT&T expresses the company's view that there is little future in mobile communications.[a]
	'When I joined Ericsson in 1984 Radio Communications was something odd happening on the outskirts of Stockholm', Kurt Hellstrom, President, Ericsson.[b]
Early-1980s	AT&T asks consultancy company McKinsey how many cellular phones there will be in the world in 2000.
	McKinsey's answer: total global market = 900,000.
	In 2000, there are about 400 million mobile phones globally (and about 180 million PCs).[c]
1992	GSM standard agreed by European standards bodies and firms.
1994	Sam Ginn quits as CEO of Baby Bell Pacific Telesis to head the company's spun-off mobile operations renamed AirTouch.
	But AirTouch's share price languishes.
1997	Demand for mobile telephony explodes globally.
1999	British Vodafone acquires AirTouch forming the largest global mobile telecoms operator (and acquires Mannesmann in 2000). Players talk of possibility of mobile replacing fixed communications.
1999	UMTS third-generation mobile standard agreed by Europe, Japan, and US, making high-speed Internet on mobile phones possible.

[a]*Financial Times*, 22 February 1999.
[b]*Financial Times*, 26 July 1999.
[c]*The Economist*, 9 October 1999.

demand thus generated. Instead, there exists a process of co-evolution, frequently involving substantial interpretive ambiguity, with consumer tastes and preferences, products and services, technology (and sometimes science), firms, and related institutions, interacting and co-evolving within the context of the various selection environments that select and reinforce some of them while rejecting others. In some cases, indeed, products and services are developed before consumer demand for them exists, with the intention of creating the very demand that the product or service is intended to satisfy. Sometimes these efforts succeed, other times they fail. Figure 2.1 depicts four possible evolutionary paths, with examples from the field of infocommunications.

Mobile cellular communications, shown in the top left-hand cell, are an example of a set of services/products/technologies that were developed well in advance of the huge global mass demand that would much later emerge. The concept of cellular communications was invented in Bell Laboratories. The first proposal to use cellular systems in the field of mobile communications in order to make most efficient use of the limited spectrum was put forward in 1947 'and discussed subsequently in a number of internal Bell Laboratories memoranda.

Development of product/ technology	Outcome	
	Successful	Unsuccessful
Before large demand	• Mobile cellular communications • E-mail	• Video-phone
After large demand	• Many	• Video-on-demand

Fig. 2.1. Evolutionary paths for co-evolving consumer demand.

These ideas formed the basis for worldwide cellular radio. The first publication was by H. J. Shulte, Jr. and W. A. Cornell [of Bell Labs] in 1960'.[68]

Only a decade later, in 1970, the first civilian standard for modern cellular telephony began to be specified in Scandinavia, leading to the Nordic Mobile Telephony (NMT) standard that was introduced in 1981. However, although demand for cellular mobile telephony in Scandinavia exceeded initial expectations, the general feeling, as indicated by the statement from Kurt Hellstrom, President of Ericsson, in Table 2.10 was that mobile communications were not particularly economically significant. Indeed, Ericsson's mobile division was very much the 'ugly sister' to the company's other division, based on the company's digital central office switch, the AXE, used at that time for fixed communications. The small market for Ericsson's equipment in mobile communications was insufficient to deflect the company from its vision for the 1980s of a merging of telecommunications and the paperless office, a vision that turned out to be incorrect. It was only in 1990 that Ericsson's new CEO, Lars Ramqvist, made mobile communications the main priority and reorganized the company accordingly.[69]

E-mail is a similar example of a service being developed well in advance of a large market demand. More accurately, e-mail was an unintended by-product of the ARPANET project and only emerged initially as a convenient way for the researchers working on this project to communicate. Once the service was available, its usefulness quickly became apparent and rapidly diffused, first amongst the ARPANET researchers, then amongst their other colleagues at their universities and research institutes, and, finally, amongst a broader community. Indeed, 'Email [introduced in 1972] quickly became the network's most popular and influential service, surpassing all expectations'.[70]

---

[68] Millman (1984: 235).

[69] 'In 1990 Ericsson got a new CEO, namely Lars Ramqvist, who came from Ericsson Radio Systems. As he came from radio communication and not switches, choosing him to be top manager was in many ways a recognition of the growing importance of wireless telecommunications to the concern as a whole, and indeed Ramqvist immediately focused Ericsson on that' (McKelvey, forthcoming: 17). See also McKelvey, Texier, and Alm (1998).

[70] Abbate (1999: 107). 'The popularity of e-mail was not foreseen by the ARPANET's planners. Roberts had not included electronic mail in the original blueprint for the network. In fact,

Table 2.11. Vision of the video-phone

Assumption	People who can hear each other will also want to see each other
1950s	Bell Labs begins work on video-telephony
1964—World Fair, New York	AT&T demonstrates the video-phone
1960s—AT&T's business plan	By 1980s, video-phone market = 1% of residential, 3% of business market
1973	AT&T withdraws video-phone after spending up to $0.5 billion
1992	AT&T reintroduces video-phone with limited success
M. Katz, former FCC Chief Economist, 1996	'Many at 1964 World Fair touted video-phone as next big thing. I think they were right, just a bit premature.'
Interpretive ambiguity	'The price is too high and people are not used to it' versus 'People often feel uncomfortable under close visual scrutiny and don't want it'

A very different co-evolutionary path for consumer tastes, preferences, services, products, and technologies is represented by the top right-hand cell. An example is the video-phone. The details of this example are summarized in Table 2.11.

By contrast, video-on-demand (VOD) represents an example of a service developed after the emergence of a large market demand for video recordings (using the VHS format). For many, VOD was the much sought-after 'killer application' for the 'multi-media' objectives that were very popular in the Telecoms Industry of the late-1980s, before the widespread adoption of the Internet. It was VOD that stimulated numerous trials around the world, and indeed the local access technology, XDSL (digital subscriber loop), designed to carry broadband signals over local copper wires, owes its origin to VOD-related research and trials. However, like the video-phone, by the dawn of the new millennium there was very little demand for VOD.

As these different evolutionary paths and examples suggest, it is necessary to analyse consumer tastes and preferences, and, therefore, market demand, as an endogenous part of the co-evolutionary process, a process that seldom settles down 'in equilibrium' for very long but which, on the contrary, is usually in a process of flux.

in 1967 he had called the ability to send messages between users not an important motivation for a network of scientific computers.... Why then was the popularity of e-mail such a surprise? One answer is that it represented a radical shift in the ARPANET's identity and purpose. The rationale for building the network had focused on providing access to computers rather than to people' (Abbate 1999: 108–9).

Scandinavia Leads

As noted earlier, work on the first standard for civilian cellular mobile communications began in Scandinavia in 1970, involving the PTTs (national network operators) and specialist equipment suppliers of Denmark, Finland, Norway, and Sweden. Although a fair amount of research was undertaken in telecoms laboratories around the world at this time, it was Scandinavia that took the lead in developing and using mobile communications. Precisely, why this commitment first occurred in Scandinavia is still unclear. Explanatory hypotheses include the large, weather-prone countries of Finland, Norway, and Sweden; the widely distributed population of these countries; the far-sighted and proactive nature of the PTTs, particularly Televerket with its own sophisticated radio research laboratories; and the cooperation of the PTTs that resulted in the first international mobile standard, NMT, which, in turn, allowed Scandinavians to use their mobile phones in any of the participating countries.

However, it is clear that whatever the explanation for Scandinavia's early entry, there were regional dynamic increasing returns that allowed Scandinavia to maintain its lead in several important areas into the new millennium. One of these areas is penetration rates where Scandinavian countries still have the highest rates in the world. Another is the Scandinavian companies, notably Ericsson and Nokia, that have come to dominate not only European but also global mobile equipment markets, overcoming the opposition of powerful rival companies such as Motorola and NEC. These dynamic increasing returns at the level of the firm have been traced by McKelvey, Texier, and Alm (1998) for the case of Ericsson and for Ericsson and Nokia by McKelvey (forthcoming).[71] The strength of these companies resulting from dynamic increasing returns has allowed them to play a highly influential role in shaping the W-CDMA (code division multiple access) standard for Third Generation mobile communications adopted by Europe and Japan and by a few operators in the US.[72]

Europe and Japan Rule—the US Lags[73]

Table 2.12 shows the diffusion of second generation mobile systems in Europe, Japan, and the US.

As can be seen from Table 2.12, in mobile communications, Europe and Japan rule while the US lags. Several qualifications, however, must be made regarding

---

[71] Ericsson was founded in 1876 to produce telephones using technologies developed by Alexander Graham Bell but not patented in Sweden.   [72] See Kano (2000).

[73] This section draws heavily on Kano (2000). The data in Table 2.12 comes from Kano (2000: table 2, fig. 2).

Table 2.12. Diffusion of second generation mobile systems in Europe, Japan, and North America

Region	Population (mn)	No. of subscribers			
		1997		1998	
		Absolute No. (mn)	Per 1000	Absolute No. (mn)	Per 1000
Western Europe	387	46.3	12.0	93.5	24.2
Japan	126	36.2	28.7	47.3	37.5
North America	300	5.5	1.8	27.2	9.1

Source: Kano (2000).

the data in Table 2.12. First, Western Europe includes countries with the highest penetration rates in the world, such as the Scandinavian countries, as well as countries with much lower rates, such as Greece and Portugal, and, therefore, is not strictly comparable with Japan and North America which is largely the US. Second, one of the reasons for the low penetration rates in the US is that the data is limited to second generation mobile. One cause of low penetration of second generation mobile in the US is the relatively high penetration rate of first generation mobile in this country. At the end of 1998, there were 50.7 million first generation subscribers in the US and Canada (using the analogue Advanced Mobile Phone Service (AMPS) standard). However, this implies a penetration of 16.9 people per 100 in North America.[74] As Kano (2000) notes, one of the reasons for the relatively low penetration rate of second generation mobile in the US was the fairly high rate of penetration of first generation.

The figures for North America (mainly the US) pose an interesting puzzle since there are not many parts of the Infocommunications Industry where the US is behind:[75] Why does the US lag in mobile communications?

Kano (2000) suggests several answers to this puzzle. The first is the lack of a single dominant standard in the US, unlike in Europe and Japan, where one standard dominated. In Europe, the dominant standard for second generation digital cellular mobile systems is Global System for Mobile Communications (GSM), which has more subscribers than any other standard in the world.[76] In Japan, the dominant second generation standard is Personal Digital

---

[74] Kano (2000: 12).

[75] For an analysis of the causes of US global leadership in the computer, software, and microprocessor semiconductor industries, see Mowery and Nelson (1999).

[76] It is interesting to note, apropos the point made earlier about dynamic increasing returns in mobile communications in Scandinavia, that the Conference on European Postal and Telecommunication Administrations (CEPT), that began work on specifying a pan-European digital standard in the early 1980s, decided in 1987/88 that GSM would become the European standard. This happened formally in 1992. GSM, however, incorporated a later version of the original Scandinavian mobile standard, NMT rather than continental European alternatives that were also proposed. This occurred even though the Scandinavian standard was narrow

Cellular (PDC). In the US, on the other hand, where the principle was accepted that the adoption of standards should be left to the market, there were three incompatible second generation standards: American National Standard Institute: ANSI-136, ANSI-95, and CDMAOne (Code Division Multiple Access). Kano suggests that the lack of a single dominant standard in the US had several other undesirable knock-on effects: some operators and users adopted a wait-and-see attitude to see which standard would dominate; there was poor geographical coverage since the standards adopted by different operators were incompatible and did not provide interoperability, in turn leading to slower user adoption of mobile services; and large-scale production of equipment and phones was frustrated by the lack of a single standard, leading to high costs.

However, Kano suggests that the US lag was not only due to the failure to provide a single dominant standard. There were several other reasons for the relatively slow diffusion of mobile communications in the US. To begin with, as already noted, the relatively rapid diffusion of first generation analogue mobile services in the US served to frustrate the take-up of second generation digital services. Furthermore, the specific characteristics of fixed telecoms services in the US also slowed the diffusion of mobile services. These included the arrangement that the called party, rather than the caller, paid for the call; the fact that a flat rate is paid for fixed local calls, making pay-per-call mobile calls relatively expensive; and the high cost of spectrum, sold by the US authorities by auction, that depleted investment resources for operators. To the extent that the flat rate for local calls was an inhibitor of the diffusion of mobile services in the US, it is ironical that the same tariffing system provided a significant boost for the diffusion of the Internet and that attempts are currently being made in Europe and Japan to imitate this system (even if at the cost of Internet congestion as a result of the low price of a scarce resource, namely Internet access.) Even more ironical is the contrast between the US's failure to generate a single dominant standard in mobile communications and the same country's superb success in generating not only US but also global standards for the Internet, partially documented earlier in this paper.

THE FOUR FORCES OF COMPETITION

Earlier in this chapter, it was stated that the technological regime and the associated learning regime constituted one of the 'engines' driving the evolution of the Telecoms Industry. The technological regime was defined in terms of the conditions surrounding the creation of knowledge and the opportunities

---

band while the continental alternatives were broadband. 'This decision on the GSM standard was very important to the Scandinavian firms Ericsson and Nokia because GSM is based on technical solutions they had already been pursuing. With this decision, they were leading the technical race, not close followers or imitators' (McKelvey and Texier 1999: 16).

Table 2.13. Competing networks and technologies in the local access market

Network/technology	Description
Copper cable/XDSL	Broadband over twisted-pair copper cable
Optical fibre	Glass optical fibre cable
Cable	Coaxial cable
Cellular mobile	Radio network, user mobile
Fixed wireless access	Radio network, user not mobile
Satellite	Communication satellites
Power Line[77]	Radio signals sent through regular electricity cables received through electricity sockets
Laser access[78]	Uses low-powered laser beams

and constraints regarding the use of that knowledge. The learning regime refers to the patterns and paths of learning that occur under the technological regime. One of the conditions 'surrounding' the creation of knowledge, which therefore is part of the technological regime, that has not been emphasized until now is the forces of competition. Schumpeter had these forces in mind when he referred to the 'fundamental impulse that sets and keeps the capitalist engine in motion', quoted at the beginning of this chapter, such as new consumer goods and new methods of production.

In the Infocommunications Industry, it is necessary to distinguish among four forces of competition: between products/services; between networks;

[77] Power Line involves the provision of data communications, including voice over the Internet and Internet access, through the electric power cables that are already connected to firms and homes, thus avoiding the need to dig up streets (as with optical fibre cables) or establish radio base stations (as with mobile or fixed radio access connections). Power Line works through radio frequencies being transmitted through electric power cables. Relatively cheap enabling technology is required at the electricity substation (serving 250 homes) and at the customer's electricity meter. The original innovation was made by an engineer working with Norweb, the electricity company that supplies the Manchester and Yorkshire areas of Britain. In the event, Norweb obtained the commanding patents for Power Line. However, Norweb, lacking the technological competencies, entered into a strategic alliance with Nortel which went on to develop the necessary technology. Although it is too soon to tell whether this new technology will be a serious substitute for other alternative local access technologies, Norweb and Nortel claim that it is significantly cheaper than the other alternatives.

[78] Laser access is a technology that was announced in March 2000 by a Seattle-based startup called TeraBeam Networks. 'The heart of TeraBeam's technology is a transmitter/receiver that is about the size of a small satellite dish and can be made for $150. Mounted near the window inside an ordinary office, it sends and receives data, as light, at speeds of up to two gigabits a second. TeraBeam does not require bulky outdoor aerials or long drawn-out negotiations with building owners to gain rooftop access, unlike similar microwave technologies. And because it does not operate in the radio spectrum auctioned by the Federal Communications Commission, it can operate without licences. [The company claims that its] optical transmission gear passes all legislated safety tests: no danger of lasers frying the eyes of the unwary'. *The Economist*, 25 March 2000, p. 89.

between technologies; and between firms. Examples of competition in each of these four areas are, respectively, competition among telex, phone (mobile and fixed), fax, and e-mail; competition among copper cable/XDSL, optical fibre, and fixed wireless for local access; competition between TDMA and CDMA in second and third generation digital mobile; and competition among AT&T, WorldCom, and Qwest. Competition in each of these areas may occur independently of competition in the other areas although sometimes there will be interdependencies between some or all of the areas.

One area where both network and technological competition is strongest is in the local access market (including Internet access). Table 2.13 shows the main competing networks and technologies providing local excess.

The creators of the knowledge embodied in networks and technologies in any of these areas have to constantly do battle to keep up with parameters defined by the alternative competing networks and technologies. For example, second generation digital mobile transmits data at 9.6 kilobits per second. Third generation digital mobile sends data at a maximum of about one megabit per second, a similar speed to ADSL. However, TeraBeam claims that its laser access technology will send data at 2 gigabits per second. This competitive environment provides a powerful context, permeated by pressures and incentives, within which knowledge-creation takes place. And a changing knowledge base provides an important engine of change. Moreover, as Schumpeter also pointed out, 'competition acts not only when in being but also when it is merely an ever-present threat. It disciplines before it attacks'.[79]

CONCLUSION

The main focus of this chapter has been on the evolution of the Telecoms Industry as it has been transformed by the Internet into the Infocommunications Industry. In analysing the causes of this evolutionary change, particular emphasis was placed on the changing technological and learning regimes, which refer essentially to the processes of knowledge-creation and use in the industry. However, while these regimes play the role of a Schumpeterian 'engine', powering the process of change, it was shown that financial markets and co-evolving consumer tastes and preferences also constitute important engines of change in the Telecoms and Infocommunications Industries. Furthermore, through the analysis of the Internet, it was seen how a paradigmatic transformation of the industry was brought about by a set of ideas, and associated technologies and services, that originally emerged from outside the Telecoms Industry, ideas that at first were vehemently rejected by the industry's technological representatives.

[79] Schumpeter (1943: 85).

A further theme in this chapter relates to the ability of the industry's participants to understand what is happening in the industry, and, in the light of their understanding, to adapt and act in what they believe is an appropriate way. At numerous junctures in the industry's evolution, it was seen that the participants confronted interpretive ambiguity, when currently available information left significant ambiguity regarding what should be inferred. Rather than 'rational' and smooth adjustment to 'given facts', we saw a mixture of responses to interpretive ambiguity. These ranged from decisive actions based on strong convictions, that subsequently turned out to be wrong (such as the original views of telecoms engineers regarding packet switching), to more hesitant and tentative plans and calculations (such as those of financial analysts). These reactions to interpretive ambiguity, were important shapers of the evolutionary process.

The evolutionary analysis of the Telecoms Industry contained both in this chapter and in Chapter 1 provides an understanding of the major forces driving the industry. How have the Big Five incumbents—AT&T, BT, Deutsche Telekom, France Telecom, and NTT—adapted to these forces? This question is examined in detail in Chapters 3–5. In Chapters 6 and 7, the adaptation of some of the main new entrants is analysed.

3

AT&T, BT, and NTT: Super-adapters or Dinosaurs in an Age of Climate Change?

In the first two chapters of this book the main forces driving the Telecoms Industry in the Internet Age were examined. This provides an analysis of the changing context in which incumbent network operators and the new entrant network operators, who entered the industry to compete with them, have had to adjust. This chapter focuses on the three incumbents from the three countries that were the first to liberalize their telecoms sectors and allow competition—AT&T, BT, and NTT from the US, UK, and Japan.

How have these three companies adjusted to the changes transforming the Telecoms Industry in the Internet Age? Have they been successful adapters, or are they dinosaurs trapped in an age of climate change? These are the questions that are tackled in this chapter.

INTRODUCTION

In 1994, I published two articles comparing AT&T, BT, and NTT.[1] The occasion provided a rare opportunity to analyse the importance of corporate strategy construction and decision-making. Although the environment of these three companies could not be 'held constant', as it would be in an ideal scientific experiment, there were important similarities in their environment. To begin with, all three companies were former national monopolist network operators, either under direct state ownership (BT and NTT) or stringent state control

---

[1] The articles were originally published as 'AT&T, BT and NTT: A Comparison of Vision, Strategy and Competence', *Telecommunications Policy*, 1994, 18(2), 137–153 and 'AT&T, BT and NTT: The Role of R&D', *Telecommunications Policy*, 1994, 18(4), 295–305. They are reprinted as Chapter 3 in M. Fransman, *Visions of Innovation: The Firm and Japan*, Oxford University Press, 1999.

(AT&T). Second, the governments of all three companies had decided at the same time, the mid-1980s, although for very different reasons, to liberalize their telecoms industries and allow new competitors. In this way they became the first countries to liberalize telecoms, blazing a trail that by the turn of the century would be followed by practically all countries in the world. All three companies, therefore, faced the challenges of having to adapt to newly competitive markets. Third, the three companies were similar in that they used the same technologies and provided the same services to the same market segments in their own countries.

With these similarities, the opportunity seemed ideal to examine the strategies and other key decisions that the companies had chosen in order, as their leaders saw it, not only to survive, but also to prosper, in their changing environment. At the same time, I thought, this 'analytic experiment' might also throw light on the debate about the relative importance of national systems and environments on the one hand, and decision-making at the firm level on the other, as determinants of both national and corporate performance.[2]

By 1994, the three companies had already experienced almost a decade of adjusting to the 'liberalization era'. How had they adjusted? More specifically, what, if any, were the differences in the visions,[3] strategies, competencies, and R&D choices made by the companies?

The answers that emerged in the 1994 articles are discussed in the following section. The beginning of a new century, however, offers a further opportunity to analyse the fortunes (and misfortunes) of these three companies, more than half a decade after the 1994 examination. How successfully have they managed to adjust, some fifteen years into the 'liberalization era'? What have their achievements and failures been, and, more importantly, how are these to be explained? To what extent are achievements and failures to be explained by choices made within these companies, to what extent by unavoidable external changes? Are the three companies super-adapters or are they dinosaurs in an age of climate change? These are the questions that are tackled in this chapter.

AT&T, BT, AND NTT COMPARED IN 1994

The most important point to emerge from the 1994 study was that the three companies had constructed significantly different 'visions' regarding their environment and what they needed to do in order to survive and prosper.[4]

---

[2] A recent important contribution to this debate is D. Mowery and R. Nelson, eds, *Sources of Industrial Leadership*, Cambridge University Press, 1999.

[3] The concept of 'vision' in these writings is elaborated upon in M. Fransman, *Visions of Innovation: The Firm and Japan*, Oxford University Press, 1999, and in http://www. TelecomVisions. com.

[4] 'Visions' are cognitive constructs created by people in order to live and act in their world. They are constructed from the person's set of beliefs about their environment (or parts of it)

Under Bob Allen, AT&T had decided that its distinctive advantage was to provide integrated information and communications solutions to customers anywhere and at any time. Its distinctive belief, however, was that the best way of achieving this goal was through *vertical integration*. More specifically, AT&T's vision was that by possessing *in-house competencies* in the three areas of telecoms networks, telecoms equipment and devices, and computers, the company would be able, better than most competitors, to provide integrated information and communications solutions. AT&T entered the 1990s with strong competencies in the first two areas—since the 1880s the company had vertically integrated networks and equipment—but not in computers. Its vision led AT&T to make a successful, but hostile, bid for the computer company, NCR.

BT, however, constructed a fundamentally different vision. Disillusioned by its own experiences in cooperating with a small group of British telecoms equipment suppliers to develop advanced technologies, BT under Iain Vallance (later Sir Iain) believed in the importance of using the market for the procurement of key network-related technologies. According to Iain Vallance, BT's distinctive competitive advantage lay not so much in its ability to create key telecoms technologies as in its superior ability to use these technologies, that in any event were increasingly available on the market, to satisfy the specific needs of specific customer segments. As a result, BT transformed its close, obligational ties with suppliers like GEC, Plessey, and STC into arm's-length, market-based relationships with many suppliers.

NTT, however, had a vision that differed from both its American and British counterparts. Based on an effective relationship going back to the late nineteenth century that NTT (in its earlier incarnations) had forged with a 'family' of four suppliers, NTT believed that it should take the lead in developing advanced telecoms technologies that were not yet available on the market. In this way NTT would be able to offer more advanced services than its competitors. Its four main suppliers were NEC, Fujitsu, Hitachi, and Oki. My 1994 papers characterized the relationship between NTT and its family of suppliers as one of 'controlled competition'. Where technologies were available on the market, NTT agreed that there was little reason to develop them internally. By the early 1990s NTT's family of suppliers had been extended to include a number of large Western companies, notably Nortel of Canada.

While AT&T's vision, therefore, was based on vertical integration, BT's relied on using the market, while NTT's favoured controlled competition. The supplier strategies that these visions led to are depicted (in pure form) in Table 3.1.

and about what needs to be done in order to achieve their purposes. Beliefs and the visions they support are sometimes wrong, resulting in 'vision failure'. In turn, this usually results in a process of 'vision revision'. Corporate 'visions' are the result of complex political, social, and ideological processes involving people in positions of power in the company sharing, debating, disputing, and negotiating their visions in order to construct a corporate vision that will guide the corporation's attempts to survive and prosper in its environment. For further details, see footnote 3.

Table 3.1. AT&T, BT, and NTT:
supplier strategies

Company	Strategy
AT&T	Vertical integration
BT	Market
NTT	Controlled competition

Source: M. Fransman (1994).

Table 3.2. AT&T, BT, and
NTT: R&D intensities, *c.* 1990

Company	R&D % sales
AT&T	7.3
BT	3.8
NTT	2.1

Source: M. Fransman, *Visions
of Innovation*, p. 104.

Table 3.3. AT&T, BT, and NTT: convergence of strategy

Company	1994	2000
AT&T	Vertical integration	Controlled competition
BT	Market	Controlled competition
NTT	Controlled competition	Controlled competition

Source: M. Fransman.

The different strategic choices made by AT&T, BT, and NTT were reflected in their different R&D intensities. AT&T was the most R&D-intensive, since, unlike the other two companies, it also produced telecoms equipment and devices, an activity requiring a good deal of R&D. NTT was the second most R&D-intensive. The reason was that although it did not itself produce telecoms equipment, it played a leading role in researching advanced technologies and equipment and undertaking much of the advanced development work. The R&D intensities are shown in Table 3.2.

AT&T, BT, AND NTT: CONVERGENCE BY 2000

By the beginning of 2000, however, a significant degree of convergence had occurred in the visions of the three companies. This convergence is summarized in Table 3.3.

As can be seen from Table 3.3, AT&T had undergone the biggest transformation. The turning point came in September 1995 when Bob Allen announced that AT&T would voluntarily split itself into three separate companies. This involved spinning off AT&T's equipment business (the former Western Electric) which became Lucent, and de-merging NCR, the computer company that AT&T acquired in a hostile takeover in 1991. Bell Laboratories, which some argued was the jewel in the AT&T crown, would go mainly to Lucent (with a relatively small contingent forming the basis for the new AT&T Laboratories). Henceforth AT&T accepted that it would have an arm's-length relationship with Lucent.

Why did AT&T abandon its vision, and the strategy that it led to, based on vertical integration? Perhaps understandably, neither Bob Allen nor AT&T ever made clear in public why their earlier vision had failed and why it was necessary to revise it. Rather, an obscuring gloss was put on the new opportunities that would become available to the three new companies.

It seems reasonable to conclude, however, that there were several factors that account for this vision failure. Most importantly, there was little evidence that there were significant synergies generated within the company as a result of the interdependencies between the company's three core areas: network operation and development, telecoms equipment and technology, and computing. To put this slightly differently, AT&T was not coming up with better integrated information and communications solutions as a result of its internal competencies in all three areas than its more narrowly focused counterparts in the US and abroad.

However, on the cost side of the equation, AT&T was having to pay a high price for having internal competencies in all three areas. First, AT&T's leadership bore the cost of complexity; that is, they had to manage three highly complex areas involving very different technologies and markets. Second, the leadership had to deal with the politics of complexity that took the form of conflicts between managers trying to further their own sectional interests.

There was also a third cost that emerged from attempting to vertically integrate the network function and the technology/equipment function. While there may have been some advantages from AT&T's network businesses being able to interact internally with the company's equipment businesses (though, as we have seen, these advantages did not give AT&T a clear competitive edge in terms of better or cheaper services), there were also important disadvantages. Specifically, the equipment businesses (that were to become part of Lucent) faced suspicion from their external customers—AT&T's competitors in the services markets—that AT&T, as the largest customer, was being given privileged access and treatment. As a result, these competitors—including the Baby Bells (the former AT&T local companies), WorldCom, MCI, and Sprint—often preferred to buy their equipment from other specialist suppliers such as the Canadian company, Nortel. By spinning off Lucent, and turning it into an independent company, Bob Allen hoped to avoid this conflict of interest. Accordingly, in 1996 Lucent was spun off. Until the beginning of 2000 it grew rapidly, so much so that its market capitalization significantly exceeded that of its erstwhile parent, AT&T.[5]

BT, however, arguably exhibited the greatest degree of continuity amongst the three companies. Its faith in using the market for its equipment needs strengthened (although, as we will see later, its R&D intensity remained practically constant throughout the period). However, although it largely ditched its former collaborating suppliers with whom it had developed complex equipment such as the System X digital switch, BT soon realized that pure market relationships were not appropriate in industries such as telecoms.

A pure market relationship involves a single transaction—such as that occurring on a spot-market—leaving buyer and seller free to go their own way thereafter. In the telecoms industry, however, buying equipment (including software) is only the first step. After the purchase not only must the equipment be maintained and repaired when it breaks down, it must also be modified and adapted so that it can be interfaced with the other elements of the constantly evolving telecoms network. Since the specialist equipment supplier has most knowledge about the equipment, this implies the necessity for the telecoms operator to have a longer term relationship with the supplier after the initial purchase transaction. Furthermore, as a result of irreducible uncertainties regarding future requirements, it is impossible for the relationship between network operator and specialist equipment supplier to be governed entirely by contract. Accordingly, it is necessary for a longer term 'obligational' relationship to be established between the two parties. It is this obligational relationship that is referred to as 'controlled competition' in Table 3.3. (The new entrants—such as WorldCom, Qwest, COLT, and Energis—also quickly learned this lesson, rapidly forming close obligational relationships with only one or two specialist suppliers.)

Like AT&T, therefore, BT also accepted that its main competence lay in operating and developing telecoms networks from elements developed by separate specialist suppliers and providing the services that customers wanted over these networks. The difference was that BT came to this realization before AT&T, although by the late 1990s the two companies had converged on this issue.

What about NTT? As we saw earlier, NTT's vision involved the belief that in order to provide superior services it should play a leading role in researching and developing complex equipment that was not already available on the market from specialist equipment suppliers. It was this belief, for example, that led NTT to take the lead in developing telecoms switches with its cooperating suppliers, the most recent being the asynchronous transfer mode (ATM) switch.[6]

---

[5] In 2000, however, Lucent's performance and share price faltered as the company suffered from its previous decision, based on a misreading of medium-term demand, to develop slower optical networks than its main rival, Nortel.

[6] For a detailed discussion of how NTT and its suppliers developed these switches, see M. Fransman, *Japan's Computer and Communications Industry*, Oxford University Press, 1995.

It is reasonable to conclude that, by the late 1990s, NTT, while not fundamentally changing its beliefs, had modified them to some extent. Although (as we shall see in the section below on R&D) NTT still firmly believes in the strategic importance of R&D,[7] it has bowed to the inevitable and accepted that in an increasing number of areas R&D should be left to specialist suppliers.

Two factors have combined to make this outcome inevitable. The first is the growing competence of the specialist suppliers and their relatively high R&D intensity compared to all the network operators (as will be shown below and in Chapter 8). The second factor is the fundamental change that has occurred in the architecture of telecoms networks with the development of Internet protocol (IP) and data networks and their reliance on smaller, decentralized units, such as servers and routers, based on computer technology. The architecture of these networks contrasts strongly with the classical, circuit-switched networks dependent on extremely complex central office switches, the new generations of which took years to develop at very high cost. In short, the architecture of the new networks has facilitated the process of vertical specialization whereby many specialist companies make rapid incremental improvements to network elements like servers, routers, and specialist software. In this situation NTT, like the other operators, has found that it can entrust more of the network-related R&D to specialist suppliers while concentrating more on R&D relating to the development of new services.

One aspect of NTT's vision, however, has remained constant since the late nineteenth century when it was originally established by the company's previous incarnation, the Ministry of Communications. This is the importance of close obligational cooperation with a small group of suppliers who, where possible, compete as well as cooperate amongst themselves. This has been referred to as 'controlled competition'.[8] However, in this connection two further points should be noted. The first is that NTT's controlled competition has evolved as a result of the two changes noted in the last paragraph. More specifically, while NTT still believes in (and excels in) close, obligational cooperation with its specialist suppliers (both Japanese and foreign), in increasing numbers of areas it feels it no longer needs to take the lead and exercise control. The second point is that all the network operators, whether incumbents or new entrants, as noted earlier, have come to realize the importance of close, obligational relationships with specialist suppliers that go beyond market contracting.

It may be concluded, therefore, that in the areas analysed in this section there has been a significant *convergence* between AT&T, BT, and NTT over the period from 1994 to 2000. But what are the most important *differences* between the three companies? In the next section we will begin to tackle this

---

[7] A belief that is also strongly evident in NTT's subsidiaries, such as its mobile subsidiary DoCoMo that does more R&D than its rivals like Vodafone.

[8] For the background on controlled competition in NTT, see M. Fransman, *Japan's Computer and Communications Industry*, Oxford University Press, 1995.

important question by analysing the evolution of corporate structure in the three companies.

AT&T, BT, AND NTT: THE EVOLUTION OF CORPORATE STRUCTURE

The evolution of the corporate structure of AT&T, BT, and NTT is shown in Fig. 3.1.[9]

Three Alternatives for Corporate Structure

The three alternatives for corporate structure are shown at the top of Fig. 3.1, under the bold two-headed arrow.

At the one extreme there is *hierarchical integration*. Under this option the 'Centre'—including the CEO, board, and top executive committees—controls the company. Although the company is divided into broad divisions, and each division further divided into businesses, the management of each division has relatively little power. Key strategic, financial, and investment decisions are made by the Centre. Greater autonomy, however, exists at the divisional and business levels regarding functions such as production, sales, and marketing.

Breakup exists at the other extreme. Here the company's leadership decides (assuming it is an internal decision and not one imposed by outside, such as by regulatory authorities) that the costs of holding the divisions together in a single company are greater than the benefits and that breakup is necessary. There are many further sub-options regarding matters such as the continuing links that will exist between the broken up companies and the ownership structure of these companies.

In the middle is what is referred to here as the *decentralization and empowerment* option. Here the company stops short of breakup but also rejects hierarchical integration. Instead an attempt is made to give significantly more autonomy than under the hierarchical integration case to the top management of the divisions. The increase in relative power of this management is frequently symbolized by the appellation CEO or President given to the leader of the division. The Centre, however, still reserves certain powers. Typically, these include the functions of overall management and financial control, appointment and remuneration of the divisional CEOs, investments over designated amounts, resolving conflicts between divisions, overall corporate strategy (although, within the context of this overall strategy, divisions are given

---

[9] This diagram is also used in Chapter 8 where the implications of this corporate reorganization for R&D are analysed.

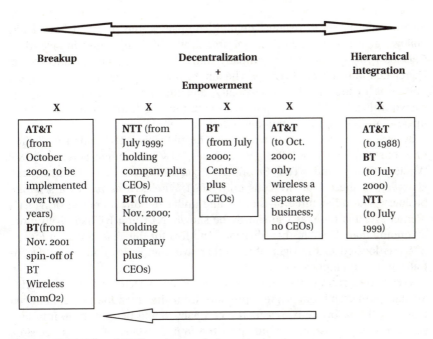

Fig. 3.1. AT&T, BT, and NTT: evolution of corporate structure.

Source: M. Fransman.

the power to make their own strategies), and maintenance of the corporation's brand and reputation.

The X's in Fig. 3.1 show the positions along the continuum of AT&T, BT, and NTT at different points in time. Crucially, the light arrow at the bottom of the figure indicates that over time all three companies have tended to move leftwards. Indeed, on 25 October 2000, AT&T was the first of the three to announce that it had finally chosen the breakup option and in November 2001 BT spun-off BT Wireless, its mobile business, which it re-named mmO2. (Significantly, the other major European incumbents—Deutsche Telekom, France Telecom, examined in Chapters 4 and 5, and Telecom Italia—refused to select the breakup option and kept their companies integrated, including their mobile businesses, with decentralization and empowerment.)

In the following section we examine the evolution of company structure by company.

AT&T.[10] In 1984 AT&T was divested, with AT&T long-lines and Bell Laboratories[11] incorporated in the new AT&T while the regional operations were

---

[10] In the Appendix further information is provided on AT&T's decision to break up into four separate companies based on a detailed interview given by CEO Michael Armstrong to *BusinessWeek*.

[11] Part of Bell Laboratories was separated and incorporated in a new laboratory, Bellcore, intended to serve the seven Baby Bells. The job of separation was undertaken by

incorporated in seven regional holding companies, the so-called Baby Bells. Following divestiture, the first major organizational change came in 1988 when CEO Bob Allen divided the company into some twenty-one business units aggregated into several divisions. The aim of this reorganization into business units was to increase focus, flexibility, and incentive. With this change, as shown in Fig. 3.1, AT&T's corporate structure moved from one of hierarchical integration to decentralization and empowerment.[12]

By early 2000, as will be shown in more detail below, AT&T had been divided into four divisions: AT&T Business Services, AT&T Consumer Services, AT&T Wireless Group, and AT&T Broadband (containing its cable TV activities). However, of these four divisions only AT&T Wireless was run as a separate business. The other three divisions came under the direct control of AT&T's 'Center'. At this time, as will shortly be shown, NTT and BT had given more autonomy than AT&T to their divisions and it is for this reason that in Fig. 3.1 AT&T is located to the right of the other two companies under the 'decentralization and empowerment' option.

AT&T's next rupture with the past occurred on 25 October 2000 when CEO Michael Armstrong, under great pressure from the stock market, announced that over the following two years the company would be broken up into four separate companies corresponding to the four divisions. With this decision, AT&T, having chosen the breakup option, moved to the extreme left of the continuum in Fig. 3.1.

Of the factors that motivated Michael Armstrong and AT&T's leadership to break the company up the most important was pressure from the stock market. Particularly troublesome for AT&T was the poor profitability of its core consumer long distance business. Poor profitability, in turn, was the inevitable consequence of increasing entry by vigorous new entrants coupled with rapidly increasing investment in new network capacity taking advantage of new technologies. This was a problem confronting not only AT&T but also its main competitors such as WorldCom. The difficulty, however, was that while earnings from the consumer long distance business were deteriorating, there were insufficient compensating increases from the new growth areas of mobile, data, and cable. The outcome was an overall deterioration in AT&T's share price. By the beginning of November 2000, just after AT&T's breakup was announced, the company's share price had tumbled 62 per cent from its peak in 1999. This meant that the company's shares were trading 20 per cent lower than when Michael Armstrong took over as CEO in 1997. The hope of AT&T's leadership was that eventually the breakup, with separate share prices and shareholdings

---

Dr Arno Penzias, Vice President of Research at Bell Labs. The bulk of the laboratory's staff, however, remained with Bell Labs.

[12] The 1988 reorganization also had major implications for the organization of R&D in Bell Labs.

for the four companies, would lead to an increase in the combined value of all of them.

A second, related factor, was the inroad into AT&T's markets made by a group of aggressive new entrants. The latter included companies such as WorldCom, Qwest, Global Crossing, Level 3, and Viatel. These companies were supported by influential stock market analysts, such as Jack Grubman, a former AT&T employee who moved to Salomon Smith Barney, one of the most influential telecoms analysts. In the views of analysts like Grubman, the new entrants were advantaged over AT&T as a result of the new technologies they were able to deploy (AT&T, like the other incumbents, was constrained by its legacy systems) as well as their corporate organizational structure. The latter made them 'F4 Firms', that is, focused, flexible, fast, and flat. Unlike AT&T, they were able to focus on chosen market segments and to develop simpler organizational structures in order to attack these markets. Support from financial markets launched a strong virtuous cycle: the share price of the new entrants increased—leading to an appreciation in the value of their 'acquisition currency'—leading to mergers and acquisitions—leading to stronger positioning as well as economies of scale and scope—leading to better performance—leading to an even stronger share price—etc. The hope of Armstrong and his colleagues was that AT&T's four companies would be able to enjoy many of the benefits of the F4 new entrants.

The decision to break AT&T up, however, implied abandoning a long cherished AT&T belief, although this was not raised by Michael Armstrong when he announced the breakup. This is the company's key belief in synergies, a belief that, as we saw, informed Bob Allen's vertical integration of networks, equipment, and computers. However, although AT&T voluntarily trivested in September 1995, spinning off Lucent and de-merging the computer company NCR, the belief in synergies continued, although in modified form. This is apparent from the following quotation from AT&T's 1999 Annual Report, issued after the company had established a 'tracking stock' for its AT&T Wireless Group in the hope that this 'will increase market awareness of the performance and value of our wireless business':

While the tracking stock will separate the economic performance of the AT&T Common Stock Group and the AT&T Wireless Group, we will maintain the advantages of doing business under a common ownership. These benefits include strategic, financial and operational synergies that might not be available if the two groups were not under common ownership. For example, the tracking stock will allow both groups to continue to benefit from the AT&T brand and from cross-marketing and bundling of services between the groups, and will also enable us to share technology, networking arrangements and business expenses.

AT&T Annual Report, 1999, p. 10.

Unfortunately, Michael Armstrong in announcing the breakup of AT&T in October 2000 did not explain what the costs of breakup would be in terms of loss of the synergies that the company had explicitly argued was a central part of its strategy. Nevertheless, he did make vague reference to 'contracts' that he intended to establish between the broken up companies so that they could still obtain some synergies.

BT. Until July 2000, BT was organized according to the hierarchical integration option. BT had two major divisions, BT UK and BT Worldwide, although these divisions contained a number of 'customer-facing' businesses (in the company's terminology).[13] These two divisions were under the direct control of BT's 'Centre'.

By the late 1990s, however, it was clear that this corporate structure was not serving BT very well. Four shortcomings were particularly important. First, it was clear that compared to its competitors BT was missing many opportunities for growth. Examples are BT's poor relative performance in wireless communications compared to Vodafone and Orange; BT's failure to capitalize on the Internet revolution compared to Freeserve and AOL in the UK and Telefonica's Terra Networks, Deutsche Telecom's T-Online, and France Telecom's Wanadoo in other European countries; and BT's slowness in taking advantage of its positioning by rapidly developing ADSL broadband services.

BT's second shortcoming was that it was unable to stem the erosion of its market share by a group of vigorous new entrants. These included companies such as COLT[14] (a company established only in 1992 by the largest US investment trust, Fidelity), Energis (the telecoms subsidiary of the English electricity company, Power Grid), WorldCom, and regional companies such as Thus (formerly Scottish Telecom, subsidiary of Scottish Power) and Atlantic Telecom (a small start-up that began in Aberdeen, Scotland and rapidly expanded its activities into the rest of the UK). All these companies shared the characteristic that they were 'F4 Firms'—focused, flexible, fast, and flat. The contrast with BT's organizational structure could not have been more pronounced.

Third, from the latter half of 2000, BT found itself being heavily penalized by a stock market that was punishing it for its weaknesses while not giving it much reward for its areas of growth and strength (precisely the same problem that AT&T also faced at the same time). Specifically, BT's eroding profitability from its core voice network dominated, in terms of stock market valuations, the improvements coming from areas like mobile and data. Like Michael Armstrong of AT&T, Sir Peter Bonfield, BT's CEO, came to the conclusion that a new corporate structure might give the company increasing financial flexibility, and therefore a higher market capitalization.

The fourth shortcoming of BT's corporate structure was that it was not well adapted for the monitoring, control, and growth of the company's increasingly important overseas investments. The division of BT into two broad units, BT UK and BT Worldwide, made it difficult for the knowledge existing in the

---

[13] See BT, *Annual Report and Form* 20-F, 1999, p. 17 and p. 94 for further details.
[14] COLT is examined in detail in Chapter 7.

company's UK businesses to be used effectively to monitor, control, and assist the growth of its foreign investments.

These were the factors that motivated Sir Peter and his colleagues in BT's Centre to announce the radical restructuring of the company in April 2000. At this time (as will be shown in more detail later) it was announced that BT would establish four new divisions: BT Wireless, Ignite, Yell, and BT Openworld. In July 2000 two further divisions were added, BT Wholesale and BT Retail. The seventh division, Concert, was a joint venture with AT&T, established in July 1998, serving international corporate customers. With this corporate reorganization, BT moved to the 'decentralization and empowerment' option. However, at this stage BT stopped short of the breakup option.

On 9 November 2000 BT made a further important announcement. Although the company would not be broken up the way AT&T had just been, BT would establish a holding company and several of its seven semi-independent companies would be partially floated, with up to a quarter of the shares being sold to outside investors.

In 2000, BT spent about £14.5 billion on acquisitions. In February 2001, BT announced a sharp fall in earnings and said it expected its total debt to amount to $43 billion. This included the cost of acquisitions and third-generation mobile licences. In March 2001, Philip Hampton, BT's new director of finance, made a public sharp criticism of the decisions that had been taken by the company's Board, alleging that a 'lack of financial discipline' had led them to make acquisitions for reasons of 'pride'. Implicitly, both the Chairman, Sir Iain Vallance, and CEO, Sir Peter Bonfield, were under attack. The attack came some days after Moody's, the credit-rating agency, had placed BT's ratings under review for possible downgrading. On 26 April 2001, BT announced that Sir Iain would retire as Chairman and be replaced by Sir Christopher Bland, chairman of the board of governors of the BBC.

Bland and Hampton then set about the task of reducing BT's indebtedness. More funding was needed. Accordingly, in May 2001 BT announced a $8.58 billion rescue rights issue, one of the largest ever undertaken on the London Stock Exchange. As a sweetener to shareholders, and to convince them that the company had re-found its strategic direction, it was also announced that BT would break the company up, separating BT Wireless (to be renamed mmO2) from the rest of the company.[15] On 26 July 2001, BT announced that its profit before tax and exceptional items had declined by 71 per cent.

After long internal debates and conflict, BT decided that mmO2 would carry £500 million in debt, far less than the £2 billion that was first talked about.

---

[15] Earlier it had been announced that BT's yellow pages business, Yell, would be floated. Subsequently, BT and AT&T agreed to dissolve their Concert joint venture. The remaining four businesses—BT Wholesale, BT Retail, Ignite, and BT Openworld—would remain as relatively autonomous businesses under the BT Board. In October 2001 Sir Peter Bonfield announced that BT had ended its process of corporate restructuring. Plans to float BT Wholesale would be abandoned.

However, this relatively light level of indebtedness was insufficient to persuade the credit-rating agencies. At the end of September 2001, Standard & Poor's assigned mmO2 a long-term corporate rating of BBB − , which is just one grade above junk status. Moody's Investors Service gave the company a Baa2 rating for both its senior unsecured bank loan and its £3.5 billion credit facility. Factors cited as influencing these ratings were mmO2's difficult competitive situation in Germany (where Deutsche Telekom and Vodafone are over-whelmingly dominant) and the company's future high capital expenditure requirements on third-generation mobile infrastructure. By the end of 2001 both Sir Iain Vallance and Sir Peter Bonfield had been forced to leave BT.

NTT. As shown in Fig. 3.1, NTT's transition from hierarchical integration to decentralization and empowerment came in July 1999 when the company was divided up into five divisions (as will be shown in more detail later) under the ultimate control of a holding company: NTT East, NTT West, NTT Communications, NTT DoCoMo, and NTT Data.

One major difference was that the NTT reorganization was influenced far more by the politics of regulation than was the case for AT&T and BT. In the US and UK the regulator was separated from the government ministry that determined the basic structure of the country's telecoms industry. In Japan, however, these two functions were concentrated in the same body, the Ministry of Posts and Telecommunications (MPT). Furthermore, there was a long-standing complicated relationship between MPT and NTT.[16] On many occasions, MPT pressed for the breakup of NTT while the latter strongly favoured remaining integrated.

In 1998 a compromise was negotiated. While NTT would not be broken up, it would establish five decentralized divisions under a holding company. Two of the divisions (NTT East and NTT West) would be regulated companies while the remaining three, operating under competitive conditions, would be more lightly regulated. Furthermore, accounting separation of the divisions would ensure that no anti-competitive cross-subsidization occurred.

In this way NTT moved from July 1999 (when the re-organization was officially implemented) from the 'hierarchical integration' to the 'decentralization and empowerment' option. Whether NTT would in any event have moved in this direction, even in the absence of regulatory pressure, is a moot point. On the one hand, for motivations similar to those that drove AT&T and BT to move in this direction, NTT's leaders may well have come to the same conclusion. On the other hand, however, there was at the time a strong feeling in NTT that there were important benefits to be gained from continuing close integration.

Conclusion. As summarized in Fig. 3.1, the evolutionary trend for AT&T, BT, and NTT was to move by mid-2000 from hierarchical integration to decentralization

---

[16] For further details, see M. Fransman, *Japan's Computer and Communications Industry,* Oxford University Press, 1995.

and empowerment. The causes of this movement, however, were to some degree similar for AT&T and BT while rather different for NTT. Specifically, stock market valuation was a more important motivator for AT&T and BT while regulatory considerations were more significant for NTT.

AT&T, BT, AND NTT: DIVISIONAL STRUCTURE, 2000

The divisional structure of the three companies in 2000 is shown in Table 3.4.

Several observations may be made about the way in which the three companies had structured their main divisions by mid-2000. The most important is that a different combination of beliefs and historical events appear to have influenced the evolution of divisional structure in the three companies. These beliefs and events will now be examined.

To some extent AT&T's divisionalization is based on the belief that divisional organization and market segment should be matched. This is most evident in AT&T Business Services and AT&T Consumer Services intended to serve the business and consumer market segments, respectively. However, this organizing principle is compromised in AT&T's other two divisions, AT&T Wireless and AT&T Broadband (containing its cable activities) that also cater to the business and consumer markets.

The belief in matching division and market segment has also informed BT's reorganization in April and July 2000. This is clearest in the case of the business segment. However, this segment is served by several of BT's divisions, raising the possibility of overlap and confusion but also, perhaps more positively, competition between the different divisions. Thus BT Retail, Ignite, and Concert (BT's joint venture with AT&T serving international business) all serve the major business markets. Furthermore, so does BT Openworld, BT's Internet service provider, that also provides ADSL services. BT Wholesale is clearly targeted at a specific market, namely the wholesale and carriers' carrier market. However, the establishment of BT Wholesale was also influenced by BT's desire to ring-fence its main regulated business in the hope that this might increase the company's overall stock market value.

The matching of corporate division and market segment seems least evident in NTT's 1999 reorganization. Here a major principle, less evident in both AT&T and BT, is the importance of geography. This is most evident in the case of NTT East and NTT West, that offer intra-prefectural services but are gradually extending their services into other markets also served by other NTT divisions. NTT Communications offers inter-prefectural services in Japan as well as global services.

A striking feature, common to all three companies, is the separation of wireless into a distinct division, namely AT&T Wireless Group (until October 2000 the only one of AT&T's divisions to be run as a separate business), BT

Table 3.4. AT&T, BT, and NTT: divisional structure, 2000

Company	Division	Description
AT&T	AT&T Business Services	Offers businesses of all sizes, long distance, local, data, Web hosting, and IP networking services. Includes AT&T Solutions, Concert, and AT&T Global Network Services (formerly IBM Global Network Services).
	AT&T Consumer Services	Provides individuals and families local, long distance, and Internet access services.
	AT&T Wireless Group	Offers consumers and businesses voice and data wireless services using mobile and fixed wireless technologies.
	AT&T Broadband	Offers cable TV, all-distance telephony, high-speed Internet access (through AT&T@Home), and digital video to homes.
BT*	BT Wholesale	Runs BT's backbone network and its local access networks; sell in wholesale, including carriers' carrier, markets.
	BT Retail	Currently BT's major link with its business and residential customers; BT Retail is a re-packager and re-seller of BT Wholesale's network services.
	BT Wireless	Includes BT Cellnet and BT's other global mobile investments.
	Ignite	Offers high-speed connections to businesses.
	Concert	BT's $10 billion joint venture with AT&T for large international businesses.
	Yell	Information based on Yellow Pages, mainly about small and medium sized businesses.
	BT Openworld	BT's Internet Service Provider, including ADSL services.
NTT	NTT East	Provides local (intra-prefectural) services in Eastern Japan.
	NTT West	Provides local (intra-prefectural) services in Western Japan.
	NTT Communications	Provides long-distance (inter-prefectural) services in Japan and all services globally.
	NTT DoCoMo	Provides mobile services in Japan and globally.
	NTT Data	Provides data services in Japan and globally.

*As noted above, from November 2001, BT was broken up into two separate companies: mmO2 (formerly BT Wireless) and BT (consisting of BT Wholesale, BT Retail, Ignite, and BT Openworld). Yell was earlier floated and Concert was broken up.

Source: M. Fransman.

Wireless/mmO2, and NTT DoCoMo. These divisions look a good deal like the 'single wireless play', Vodafone. So much for 'fixed-wireless integration'!

Apart from beliefs in organizing principles, historical events have also played an important role in determining divisional structure. This is clearly the case with AT&T Broadband, set up as a separate division after the company's late move into cable TV as a way of developing a local access network. (In March 1999 AT&T spent $50 billion in acquiring TCI, the second largest cable company in the US, and in May the same year it spent another $57 billion on its merger with Media One.) Similarly, BT's attempts to deal with the impact of regulation on its market capitalization led it to establish BT Wholesale. The establishment of NTT DoCoMo and NTT Data was also influenced by the regulator wanting to limit the possibility for cross-subsidization and ensure fair competition.

It may be concluded, therefore, that while beliefs in corporate organization have certainly played an important role in determining the divisional structure of the three companies, historically and regulatory contingent events have also been significant.

AT&T, BT, AND NTT: PERFORMANCE IN THE NEW GROWTH AREAS

Until this point we have been largely concerned with the vision, strategy, and structure of AT&T, BT, and NTT. But how well have they performed? This question will be tackled by an overall evaluation of their performance in the three new growth areas of the Telecoms Industry: mobile, Internet, and globalization.

Performance in Mobile

The performance of the three companies in mobile is summarized in Table 3.5.

In Table 3.5 performance is measured against the other leading players in the national market. According to this measure, AT&T's performance in mobile is 'medium'. Further evidence supporting this conclusion is provided in Table 3.6.

AT&T was relatively slow to enter the field of mobile communications and when it did it was the merger route to entry that was chosen. It was only in 1993 that AT&T merged with McCaw Cellular Communications, set up by Craig McCaw, in an agreement worth $12.6 billion.[17] But by mid-2000 AT&T Wireless

---

[17] AT&T's main long-distance competitor, the new entrant MCI-WorldCom, was even slower to realize the importance of mobile communications. For a long time arguing that mobile was not important for WorldCom, Bernie Ebbers, the company's CEO, only changed his beliefs in 1999. This led WorldCom to attempt to enter the mobile market through the acquisition of Sprint. This acquisition was thwarted, however, by both the US and European regulatory authorities.

Table 3.5. AT&T, BT, and NTT: performance in mobile

Company	Performance	Origin	National comparators
AT&T	Medium	Acquisition (McCaw)	Sprint, Verizon Wireless (AirTouch, Bell Atlantic, GTE), Cingular (SBC/Bell South)
BT	Low	Start-up (BT Cellnet)	Vodafone, Orange
NTT	High	Start-up (DoCoMo)	J-Phone, KDDI (Tu Ka, IDO)

Source: M. Fransman.

Table 3.6. The US mobile market

Company	Subscribers (in millions)	Revenues (in millions)	Net income/(loss) (in millions)
Verizon Wireless	25.4	$13,543	$1,409
AT&T Wireless	11.7	$7,627	($405)
Sprint PCS	7.4	$3,180	($2,502)
Nextel	5.6	$3,326	($1,530)

Financial information is from 1999; subscriber figures are from June 2000.

Source: *Business Week*, 23 October 2000.

was only the third player in the US mobile market. At this time AT&T Wireless had less than half the subscribers of the leader, Verizon Wireless (a joint venture formed by Vodafone/AirTouch, Bell Atlantic, and GTE). The second player (not shown in Table 3.6) is a joint venture, Cingular, formed by two of the former Baby Bells, SBC and Bell South. AT&T Wireless' revenue was just over half that of Verizon Wireless.

BT has also performed relatively[18] poorly in mobile communications. Relative to the UK and global market leader, Vodafone, BT's performance has been extremely poor. The relatively poor performance of BT Cellnet is shown in Table 3.7.

Several important points emerge from Table 3.7. The first, comparing the two UK companies, Vodafone and BT Cellnet, is their similarity in UK market share but their large difference in market capitalization. Why such a large discrepancy? The answer to this apparent puzzle is that unlike BT Cellnet, Vodafone has significant global mobile assets and, indeed, is the best globally positioned mobile operator in the world.[19] Under Sir Gerald Whent, from 1988, Vodafone

[18] In absolute terms, however, both AT&T and BT have performed well with rapidly growing revenues reflecting the fast global growth in demand for mobile services.

[19] The puzzle as to why it was Vodafone, rather than any of its counterparts around the world, that established such a strong global position is explained in M. Fransman, Vodafone's Rise as a Global Mobile Operator (mimeo).

Table 3.7. NTT DoCoMo, Vodafone, and BT Cellnet: market share and market capitalization

	NTT DoCoMo	Vodafone	BT Cellnet
Domestic market share	56%	37%	30%
Market capitalization ($bn, Feb. 2000)	336.3	339.7	20.3

Source: Calculated from *Financial Times*, February 2000.

made an earlier and more substantial commitment to a global extension of its domestic activities than any of its counterparts. Indeed, by the end of 1999, after Vodafone's acquisition of the US company AirTouch, but before its acquisition of the German firm Mannesmann, more than 50 per cent of Vodafone's stock market value was generated outside both the UK and the US.[20] Furthermore, in the UK BT Cellnet has been outperformed by Orange (which in 2000 was acquired by France Telecom). Orange only entered the UK market in 1993 (together with One-2-One) and did not enjoy the benefits of duopoly as Cellnet and Vodafone enjoyed from 1985 to 1993.

Second, at this time NTT DoCoMo and Vodafone had about the same market capitalization even though NTT DoCoMo had almost 60 per cent of the domestic market compared to Vodafone's almost 40 per cent. Once again this is explained by Vodafone's superior geographical positioning.

Third, Vodafone's market capitalization was seventeen times greater than that of BT Cellnet. It is for all these reasons that BT Cellnet's performance in mobile has been judged to be 'low'.

Table 3.7 also provides evidence for judging NTT DoCoMo's performance to be 'high'. In addition to its large share of the Japanese market, NTT DoCoMo has also proved itself to be an innovative global leader with its development of its i-mode mobile Internet access service.[21] Indeed, NTT DoCoMo has received substantial international recognition for this service that has led the world in mobile data services. For these reasons the company's performance in mobile has been judged to be 'high'.

Performance in Internet-related Services

The performance of the three companies in Internet-related services is summarized in Table 3.8.

One of the more important observations to be made about the new Info-communications Industry that emerged in the mid-1990s with the mass

---

[20] Based on a sum-of-the-parts valuation.

[21] Why NTT DoCoMo succeeded with i-mode while Europe failed with WAP (a protocol used to achieve the same result as i-mode, namely accessing web sites through mobile handsets) is explained in the final chapter of this book.

Table **3.8**. AT&T, BT, and NTT: performance in
Internet-related services

Company	Performance	National comparators
AT&T	Medium/low	AOL, Yahoo, WorldCom/UUNet
BT	Medium/low	Freeserve
NTT	High/medium	Niftyserve

Source: M. Fransman.

adoption of the Internet and the transformation of the old Telecoms Industry[22] relates to the pioneering role played by new entrants. For example, in the layers of the Infocommunications Industry providing functions such as connectivity and navigation it is new entrants rather than the incumbents of the telecoms, computer, and software sectors that have dominated. Several examples of this are shown in Table 3.8, which analyses the performance of AT&T, BT, and NTT relative to dominant competitors in the domestic market in Internet-related services.

In the US, AT&T has made little headway in Internet-related services. Rather, new entrants such as WorldCom/UUNet, AOL, and Yahoo have dominated the connectivity (Internet access provision and Internet service provision) and navigation functions. WorldCom/UUNet is the largest Internet backbone service provider in the country; AOL is the largest Internet service provider; and Yahoo runs the most successful portal. In the UK, BT has similarly failed to establish a strong position in these areas and has been significantly outperformed by the local Freeserve (established by an electronics retailer) and US subsidiaries like AOL and Lycos. Perhaps BT's most significant 'vision failure' in the Internet area was its failure to see the significance of free Internet access (where the subscriber paid only for the local call). It was this innovation that propelled Freeserve into the number one position in the UK. In Japan, NTT has, compared to AT&T and BT, been more successful with its OCN (open computer network) business which, however, has not been able to challenge Niftyserve for dominance. NTT, however, is Japan's biggest carrier of Internet traffic. (Niftyserve, established originally by the computer company Fujitsu as a network for its customers and adapted to the Internet, is one of the few examples of an incumbent succeeding in these layers of the industry.)

AT&T, BT, and NTT, however, are making significant efforts to strengthen their positions in Internet-related services. One of the main aims of AT&T's investment of some $110 billion in cable TV networks is to improve the company's position in broadband Internet access. One of BT's seven new

---

[22] For further details regarding the Infocommunications Industry see Chapter 2 and http://www.TelecomVisions.com.

Table 3.9. AT&T, BT, and NTT: performance in
globalization

Company	Performance	National comparators
AT&T	Low	WorldCom, Qwest, Global Crossing, GTS, Level 3
BT	Medium	Vodafone, COLT
NTT	Only began 1999	—

Source: M. Fransman.

businesses is BT Openworld, its Internet service provider that will also offer ADSL Internet access. While OCN was located in NTT Communications in the July 1999 reorganization, several other NTT businesses, such as NTT East and NTT West, are attempting to compete in Internet-related services. From a different point of entry, NTT DoCoMo has become the largest mobile Internet access provider, not only in Japan but also in the world (as shown in detail in Chapter 9).

Nevertheless, despite these efforts, it is reasonable to conclude that the performance of the three companies in Internet-related services has been moderate, with NTT performing somewhat better than the other two companies.

Performance in Globalization

An important potential source of growth for all companies is provided by the global market outside their own home country. At the present time, AT&T, BT, and NTT have all stated that globalization is an important part of their growth strategy. The performance of the three companies, however, is mixed. This performance is summarized in Table 3.9.

In 1997 AT&T reiterated the importance of globalization as one component in its 'three-part strategy', emphasizing its goal to 'broaden and deepen our global reach'.[23] How much progress has the company made in achieving this strategic objective? In 1992 AT&T reported that 8.7 per cent of its total revenue came from 'operations located in other countries'.[24] This figure, however, included the revenues of NCR, a relatively globalized computer company that merged with AT&T on 19 September 1991. Although the figures are not strictly comparable, AT&T announced in 1999—after the de-merger of NCR and Lucent had been announced in September 1995—that 'international operations and ventures'[25]

---

[23] AT&T Midyear Report, 30 June 1997, p. 1.   [24] AT&T, Annual Report, 1992, p. 22.
[25] 'International operations and ventures' include: AT&T's consolidated foreign operations, international carrier services, international online services, and the earnings or losses of AT&T's non-consolidated international joint ventures and alliances. However, bilateral international long distance traffic is excluded. AT&T Annual Report, 1999, p. 24.

Table 3.10. TMA Annual Survey—question: how
do TMA members rate UK operators for quality of
service?

Operator	1998 score	1997 score
COLT	7.24	7.55
WorldCom	7.00	6.81
Kingston	6.85	6.87
Energis	6.70	6.47
BT	6.38	6.72
AT&T	6.18	6.42
Cable & Wireless	6.11	6.12

Source: TMA Annual Membership Survey.

accounted for 1.97 per cent of the company's total revenue. In 2000, 5 per cent
of WorldCom's revenue came from outside the US.

Further evidence of AT&T's relatively weak global performance comes from
the UK, one of the company's priority global markets when it embarked on its
strategy of globalization. The evidence is provided in Table 3.10 which shows
the results of the authoritative annual survey of the (British) Telecommunications Managers' Association (TMA), the organization that represents professionals managing telecommunications departments for most of the largest
companies in the UK. In view of the significant proportion of the largest
companies in the UK (the largest users of telecoms services) represented in the
TMA, this survey may be taken as one of be the best assessments available of
the relative performance of the various telecoms operators competing in the UK
market.

As Table 3.10 shows, AT&T performed poorly in the UK, relative not only to
BT, but also to the new entrants—COLT, WorldCom, Kingston, and Energis—
that outperformed both the American and the British incumbents. COLT and
WorldCom are also American companies that began activities in the UK in the
early 1990s, later than AT&T. Shortly after this survey came out, AT&T made the
strategic decision to effectively end its contesting of the British market and
rather enter into an alliance with BT.

BT is more globalized than AT&T in terms of the proportion of its assets held
outside its home country, although this to some extent is an indication of the
relatively small size of the UK market relative to the US. However, in the 2000
financial year, 'approximately 95 per cent of the [BT] group's turnover was
generated by operations in the UK, compared to 96 per cent in the 1999 and 1998
financial years'.[26]

BT's globalization strategy has not been particularly successful. A major blow
came in July 1997 when BT's merger with MCI (then the second-largest long

[26] BT, Annual Report, 2000, p. 39.

distance carrier in the US) was thwarted by WorldCom that succeeded in acquiring MCI. One year later, in July 1998, BT and AT&T announced the formation of a $10 billion joint venture, named Concert, to serve multinational business customers, international carriers, and internet service providers worldwide. Concert was launched on 5 January 2000.[27] Unlike Deutsche Telekom, therefore, which in 2000 launched a bid for the US mobile company, Voice Stream, BT has decided to rely on its strategic partnership with AT&T in the US market.[28] BT also has a relatively weak position in Asia. Its most substantial investment in this region was its acquisition of a 15 per cent share in Japan Telecom completed in August 1999.[29] AT&T also has a similar share in Japan Telecom.

BT's main strength outside the UK has been in Europe. Here the company's strategy has been to initially take a minority holding in a consortium led by local companies, usually not with a telecoms background, hoping to become a major challenger to the country's incumbent. Relying on its telecoms competencies, BT's strategy is to gradually increase its holdings, eventually becoming the majority shareholder. For example, this is what BT has done with the German company Viag (involved both in fixed and mobile communications), in 2000, increasing its share from 45 per cent to 90 per cent.

However, the shortcomings of a strategy based on initial minority shareholdings have been criticized. Many of the weaknesses were revealed in October/November 2000 when a serious dispute erupted in the Italian mobile company, Blu, the fourth-largest in Italy. As a minority partner with a 21 per cent shareholding BT had agreed with its non-telecoms partners to assume a majority shareholding. It appears, however, that BT was unable to fulfil its commitment largely because of its growing indebtedness, the result of high investment and third-generation mobile licence costs. This, however, led to Blu's withdrawal from the auction for Italian third-generation mobile licences at the earliest stages of the auction. In turn, this resulted in major political fallout in Italy since there was only one more company bidding than there were licences. Accordingly, the licences had to be awarded to the remaining bidders at a price substantially lower than in Britain and Germany.

NTT is in a separate category regarding globalization. The reason is that under regulatory conditions imposed by the Japanese Ministry of Posts and Telecommunications, NTT has only been allowed to operate globally in an unrestricted way from July 1999. Globalization, however, remains a high priority for NTT and its subsidiaries NTT DoCoMo and NTT Communications are making rapid progress. Notable deals have been NTT DoCoMo's acquisition of a 20 per cent stake in Hutchison's Hong Kong mobile company, a 15 per cent

---

[27] AT&T, Annual Report, 1999, p. 28.
[28] In 2001, as part of its restructuring, BT abandoned its Concert joint venture with AT&T.
[29] As part of its restructuring in 2001, BT sold its interests in Japan Telecom and its mobile subsidiary, J-Phone, to Vodafone.

interest in the Dutch incumbent's mobile subsidiary, KPN Mobile, and a 15 per cent holding in the third-generation mobile company established in the UK by Hutchison and TIW.[30] NTT Communications has acquired a majority stake in Verio, one of the largest web-hosting companies in the US, as a result of its $5.5 billion investment in the company. However, important as these moves are, insufficient time has passed in order to be able to compare NTT's success in globalization with that of AT&T and BT.

It is clear, however, that both AT&T and BT have not performed as well as domestic rivals in their efforts to globalize. Some of AT&T's domestic competitors who have been highly successful in globalizing include WorldCom (that many analysts argue is the best globally-positioned of the fixed telecoms carriers); Qwest (that has a successful pan-European joint venture with the Dutch incumbent, KPN-Qwest); Global Crossing (that has constructed a global high capacity network in a short period of time); and COLT (that has been the most successful telecoms company on the London Stock Exchange and is owned by the US mutual fund company, Fidelity).[31] Vodafone is BT's most successful domestic rival in the globalization stakes, although both companies started in mobile at the same time.

It may be concluded, therefore, that, while AT&T's performance in globalization has been low, BT's has been moderate (although as noted earlier the relatively small size of the British market and the proximity of European markets provides part of the explanation for BT's moderate performance in globalization). As also noted, it is too early to be able to judge NTT's success in globalization (including that of its subsidiaries).

AT&T, BT, AND NTT: R&D

As was shown earlier, the R&D behaviour of AT&T, BT, and NTT differed significantly in 1994, reflecting their strategic choice of vertical integration, the market, and controlled competition, respectively. How does their behaviour compare in 2000? Data on the three companies' R&D is shown in Table 3.11.

The most notable aspect of Table 3.11 is the change in AT&T's R&D intensity. By far the most R&D-intensive of the three companies in 1987, AT&T became by far the least intensive by 1999. The main reason for this significant reduction, as noted earlier, is the spinning off of Lucent in September 1995 since telecoms equipment is a far more R&D-intensive activity than network operation and service development. Equipment suppliers such as Lucent, Nortel, and Ericsson spend about 15 per cent of their sales on R&D. However, AT&T's R&D intensity also reflects the company's strategic choice regarding the role of R&D. Evidence

---

[30] NTT DoCoMo's globalization strategy is analysed in Chapter 9.
[31] COLT's pan-European activities are examined in detail in Chapter 7.

Table **3.11.** AT&T, BT, and NTT: R&D, 1987 and 1999

Company	1999 R&D spend ($'000)	Sales ($m)	R&D % sales	R&D per employee ($'000)
AT&T				
1999	550,000	62,391	0.9	2.3
1987			7.3	
BT				
1999	556,037	30,163	1.8	2.6
1987			2.1	
NTT				
1999	3,729,910	95,061	3.9	10.3
1987			3.8	

Source: 1999: Financial Times R&D Scoreboard, 19 September 2000; 1987: M. Fransman, *Visions of Innovation*, Oxford University Press, 1999, p. 103.

supporting this contention comes from the significantly higher R&D intensity of BT and NTT, with BT spending about the same in absolute terms as AT&T, but having twice the latter's R&D intensity. Significantly, AT&T's new entrant competitors—such as WorldCom, Qwest, and Global Crossing—spend an insignificant amount on R&D, preferring to outsource most of their R&D requirements to equipment suppliers such as Nortel and Cisco. It is possible that, viewing the performance of these companies, AT&T decided to move in the same direction while not going as far. AT&T Laboratories—the new R&D laboratory that replaced Bell Laboratories, most of which went to Lucent—continues to be viewed as important by AT&T.

R&D intensity in both BT and NTT has remained more consistent over the period. However, NTT's R&D intensity is almost twice as high as BT's. It is noticeable that NTT in absolute terms spends about seven times as much on R&D as the other two companies. This is a reflection of the fact that NTT continues to believe that R&D has a key role to play as a determinant of its competitiveness and broader contribution to Japanese society. NTT is the only one of the three incumbents to be obliged by law to undertake R&D in the society's interests, although France Telecom has a similar arrangement. However, although high compared to AT&T and BT, NTT's R&D intensity is not high compared to incumbents Telia from Sweden (3.3 per cent) and Telstra from Australia (3.1 per cent). Furthermore, NTT's R&D intensity is not high compared to the average figure for several industries not normally considered as 'high-tech': Vehicles (4.2 per cent), Leisure and Hotels (3.2 per cent), and Building Materials (3.0 per cent).[32] Significantly, NTT DoCoMo, the world's leading mobile Internet service provider with its highly successful i-mode service, attributes much of its success in Japan to the role played by R&D and is basing its global strategy largely on its R&D-based global technology lead.

[32] *Financial Times*, R&D Scoreboard, 1999, see also http://www.TelecomVisions.com.

Table 3.12. AT&T, BT, and NTT: common sources of pain in 2000

Falling revenue and profit from core voice services
Time before new growth areas contribute compensating revenue and
profit
Vulnerability of legacy networks to attack by specialist equipment
supplier-supported new entrants
High investment costs
Falling share price

Source: M. Fransman.

AT&T, BT, AND NTT: COMMON SOURCES OF PAIN AT THE START OF THE NEW MILLENNIUM

AT&T, BT, and NTT have faced common sources of pain. Some of the most important are shown in Table 3.12.

All three companies have faced falling revenue and profit from their core voice services. AT&T has been particularly hard-hit by falling revenues from consumer long-distance services that declined 11 per cent in 2000. This has been the cause of a significant decline in the company's share price.[33] BT's turnover from fixed-network calls, its most important single source of turnover, declined from £6.04 billion in 1996 to £5.91 billion in 2000.[34] Similarly, NTT sales from wireline services fell from 6.14 billion yen in 1997 to 5.75 billion in 1999.[35]

The problem for the three companies, however, is that the new growth areas with which they are involved in the Infocommunications Industry are not yet contributing sufficient revenue and profit to compensate for the declining core services. AT&T's situation, typical of all three, is shown in Table 3.13.

As is clear from Table 3.13, revenue from AT&T Wireless and AT&T Broadband (established as a result of acquisitions costing about $110 billion) is still relatively small. Although revenues from these businesses are growing fairly rapidly, they are insufficient to compensate for the falling tariffs that are hitting AT&T Business and AT&T Consumer. BT's similar position is shown in Table 3.14.

As shown in Table 3.14, turnover from BT Wireless, Ignite, and BT Openworld increased fairly rapidly between 1999 and 2000. However, these businesses' contribution to operating profits was less significant. Indeed, both Ignite and BT Openworld made losses in both years. Table 3.15 shows that NTT's operating income from wireline services has been falling rapidly. Although operating income from wireless services has been increasing rapidly, this has been insufficient to prevent a sharp decline in overall operating income for the three segments.

---

[33] See, for example, *Business Week*, 6 November 2000, pp. 106–7.
[34] BT, Annual Report 2000, p. 9.   [35] NTT, Annual Report 1999, p. 52.

Table 3.13. AT&T: revenue and earnings from its four businesses, 2000

Business	Revenue and earnings
AT&T Business	Revenue: $28.4 bn
	Earnings: $10.3 bn
AT&T Consumer	Revenue: $19.7 bn
	Earnings: $8.1 bn
AT&T Wireless	Revenue: $9.6 bn
	Earnings: $1.6 bn
AT&T Broadband	Revenue: $9.3 bn
	Earnings: $2.3 bn

Source: *Financial Times*, 26 October 2000.

Table 3.14. BT: turnover and operating profit from its seven new businesses

Business	Description	Total turnover		Total operating profit (loss)*	
		(£ bn)	(£ bn)	(£ bn)	(£ bn)
UK Wholesale & UK Retail	UK Wholesale: UK copper access network, core circuit switched network, and local exchanges UK Retail: Account-managed and packaged solutions for business and residential customers	12.6	11.7	3.4	3.4
Ignite	Broadband internet protocol (IP) business	3.2	2.1	(0.5)	(0.2)
BT Openworld	Mass-market internet business	0.1	0.0	(0.3)	(0.1)
BT Wireless	International mobile business	4.5	2.7	0.2	0.1
Yell	International directories and e-commerce business	0.6	0.5	0.2	0.2
Concert	Joint venture with AT&T serving international companies	1.9	1.9	0.3	0.3
Eliminations and other		(1.0)	(0.7)	0.2	(0.1)
Total		21.9	18.2	3.5	3.6

*Before goodwill amortization and exceptional items.

Source: BT Annual Report 2000, pp. 9 and 113.

Table 3.15. NTT: segment operating income

Segment	1997 (Yen bn)	1998 (Yen bn)	1999 (Yen bn)
Wireline services	562.7	467.2	83.3
Wireless services	106.9	327.2	467.2
Data communication services	41.4	43.2	29.6
Other	34.1	2.6	52.5
Total	745.2	840.2	633.0

Source: NTT Annual Report 1999, p. 52.

A further source of pain for all three companies has come from their vulnerability resulting from their legacy networks. Although these networks have already been largely depreciated and their continuing usefulness (enhanced by upgrading technologies such as ADSL) means they are not worth scrapping, the downside is that they are not as efficient as the networks, embodying only the latest technologies, established by their new entrant competitors. The latter include companies like Qwest, Level 3, and Global Crossing in the US, and COLT and Energis in the UK.

A further source of vulnerability for AT&T, BT, and NTT has resulted from their high investment costs. In AT&T's case an important investment cost has arisen from the company's strategic decision to acquire its own local network (lacking since its separation from the Bell regional holding companies at the time of divestiture). Armstrong's decision to gain local access through the acquisition of the cable TV companies, TCI and Media One, has cost the company around $110 billion. In addition, AT&T has had to upgrade its cable networks to provide broadband communications, to upgrade its existing legacy networks, and to develop new networks such as its mobile network and its international IP network jointly owned with BT (until the ending of their Concert joint venture in 2001). BT and NTT already had their own local networks. However, both of them have made heavy investment commitments in connection with their globalization and have also had to upgrade their domestic networks and build new ones. BT, in particular, has been hard pressed by the high cost of purchasing auctioned third-generation mobile licences in Europe. This has resulted in a significant rise in the company's indebtedness, that in turn has prompted the decision, announced on 9 November 2000 to sell-off one quarter of the shares of BT Wireless and Yell.

The result of these woes has been a sharp fall in the share price of all three companies, accentuated by a generalized global bear market in the technology and telecoms sector that began in March 2000. The fall in the price of AT&T's and BT's shares is shown in Table 3.16, that also compares several other telecoms companies. NTT's share price also began falling from mid-1999.[36]

---

[36] Further details are provided in Chapter 1 on the Telecoms Boom and Bust.

Table 3.16. Falling telecoms share price
(year to mid-October 2000)

Company	Percentage fall (%)
BT	−52
Deutsche Telekom	−48
AT&T	−47
France Telecom	−21
Vodafone	−19
Telefonica	−9

Source: Bloomberg Financial Markets.

Table 3.16 also shows that by October 2000 not all the telecoms incumbents were equally hard hit. Specifically, Telefonica and France Telecom suffered less than AT&T, BT, and Deutsche Telekom, largely because they had been more successful in the telecoms growth areas in their own countries. France Telecom, for example, unlike AT&T and BT, is the front runner in both mobile and Internet access in its country. Telefonica has been extremely successful with its Internet company, Terra Networks that in 2000 merged with the US Internet service provider, Lycos. Vodafone is also an interesting contrast to AT&T and BT. While the latter two companies have suffered from slow-down and intense competition in the 'traditional' areas of voice over fixed networks, Vodafone, as a 'single play' company, is only involved in the rapidly growing area of mobile communications.[37]

It may be concluded, therefore, that AT&T, BT, and NTT have suffered similar pains. In the following section a more detailed analysis is provided of the structural forces that have been operating 'behind the backs' of all telecoms operators, forces that have inflicted significant pain on most of them.

FORCES OPERATING BEHIND THE BACKS OF THE TELECOMS OPERATORS

The main structural forces that have been operating 'behind the backs' of the telecoms operators are shown in Table 3.17.

Table 3.17 shows the six interrelated forces with which all telecoms operators have to deal. The first is a radical change in demand. Specifically, the demand for fixed voice services—more graphically described as 'plain old telephony'—has declined significantly. At the same time demand has rapidly increased for a

---

[37] Significant uncertainty surrounds Vodafone, however, regarding future revenue and profitability in third-generation mobile where the company has committed itself to high network and auctioned licence fee costs.

Table 3.17. The forces operating behind the backs of the telecoms operators

The force	Explanation
1. Radical demand changes	From the mid-1990s, huge increases in demand particularly from the Internet and for mobile communications.
2. Explosion in capacity, increasing supply	Examples: advances in optical fibre technology (e.g. DWDM), leading to plunging marginal cost of carrying information approaching zero, leading to rapid fall in price of any-distance voice and data; advances in bandwidth compression and device technologies; IP.
3. Technological competition	Competition between technologies, e.g. optical fibre vs copper cable (ADSL) vs fixed wireless access vs cable TV (coaxial cable) vs cellular mobile vs power line vs laser line.
4. Capital markets (debt and equity)	Have poured hundreds of billions of pounds into new telecoms networks and complementary resources (physical and human) embodying new technologies. This has increased capacity (see 2).
5. Growing inter-firm competition	Entry by large number of new entrants, many with no background in telecoms; made possible by vertical specialization between equipment and network layers of the industry; leading to: dramatic *lowering* of entry barriers, great increase in *intensity* of competition; *falling* prices.
6. Economics of the infocom industry	High fixed/sunk costs; low marginal costs, economies of scale, scramble for traffic/customers to spread fixed costs, lower average costs, become low-cost competitor; low marginal costs lead to low price; also economies of scope.

Source: M. Fransman.

range of new services, particularly mobile, data, and Internet-related. As shown in Tables 3.14–3.16, demand for mobile services has become an extremely important source of revenue for AT&T, BT, and NTT. (With this in mind, it is all the more ironical to recall that in the early 1980s McKinsey advised AT&T that in 2000 there would be about 900,000 mobile phones in existence globally. In fact, there were about 400 million.)[38]

---

[38] See M. Fransman, 'Evolution of the Telecoms Industry into the Internet Age', Articles Zone, http://www.TelecomVisions.com.

Second, there has been an explosion of carrying capacity, increasing overall supply. The increase in capacity has come from two sources: increasing investment in networks, and improvements in technology. An example of the latter is DWDM which uses different coloured wavelengths to increase the amount of information sent along a single optical fibre.

As rapidly as demand is increasing, boosted by the Internet and mobile communications, some calculations suggest that supply is increasing even faster. One calculation is given in Table 3.18 which shows demand and supply forecasts for the US and Europe for 1999–2004.

If supply is increasing more rapidly than demand, this should be reflected in falling prices. Indeed, there is evidence to suggest that this is precisely what is happening. Some of this evidence is presented in Table 3.19 which shows a significant fall in the prices of trans-Atlantic and pan-European circuits over the years 1997–2001.

In fact, so many bits of information can now be transmitted along a fibre that the marginal cost of doing so (i.e. the cost of sending an additional bit) is approaching zero. Since under competitive conditions the price tends towards the marginal cost, this implies, in the presence of sufficient competition, strong

Table 3.18. Demand and supply of optical fibre, 1999–2004

Demand and supply	Compound annual growth rate (1999–2004) (%)
US	
Demand (effective peak demand, terabits per second)	62
Supply (lit capacity, terabits per second)	79
Europe	
Demand (effective peak demand, terabits per second)	66
Supply (lit capacity, terabits per second)	101

Source: Renaissance Strategy, 2000 (modest supply growth scenario).

Table 3.19. Falling cost of trans-Atlantic and pan-European circuits

Product	Price
Transatlantic circuit leased for 15 years	
1997	$16 million
2001	$0.85 million
Pan-European circuit leased for 10 years	
1998	$18 million
2000	$6 million

Source: *Financial Times*, 11 September 2000.

pressure for the price of carrying information to also approach zero. It is this simple fact that lies behind the rapidly falling prices and revenue that, as we have seen, have been causing so much trouble for AT&T, BT, and NTT.

A third significant force operating 'behind the backs' of the telecoms operators is technological competition. Reference has already been made to individual technological breakthroughs such as DWDM. Technological competition, however, refers to the process whereby improvements in one technology are brought about *in response to* improvements in other technologies that can deliver the same outcomes. Technological competition often comes about because groups of people (in firms and perhaps universities) have a vested interest in the development of a technology. A good example is ADSL, that has been referred to as the 'incumbent's technology' because it is particularly important for incumbent telecoms operators who have already invested in copper cables connecting to final users. But incumbents and other competing operators will also use optical fibre cables that provide an alternative way of providing local access. The R&D engineers developing ADSL technology, therefore, have to at least keep up with, if not beat, the performance parameters being achieved by optical fibre cable. The same process extends to the other competing local access technologies, that include cable TV (coaxial cable), fixed wireless access, cellular mobile, power line (that uses already-installed electricity cables), and laser line (that uses lasers directly for local access).[39] The R&D engineers with a vested interest in these technologies have to do precisely the same thing. Indeed, there must be very few other examples, either from other existing industries or from history, where so many technologies have competed as is presently the case in the Telecoms Industry. This produces a powerful force in the industry, leading to what Schumpeter called waves of creation–destruction.

A fourth force has been generated by capital markets (both debt and equity). Responding to hoped-for high earnings in the wake of Internet- and mobile-generated increases in demand, capital markets have poured hundreds of billions of dollars into new telecoms networks and companies. This investment, in turn, has fuelled the increases in capacity already referred to. Furthermore, capital markets have generated other dynamics that have produced forces operating 'behind the backs' of the operators. For example, they have facilitated the entry of new telecoms operators who, as we have seen, have sometimes been able to establish a technological edge over the incumbents, constrained by their legacy networks. Capital markets have also enabled successful new entrants—Vodafone and WorldCom are amongst the most dramatic examples—to grow rapidly. This has occurred as enthusiastic investors have pushed up the share price and market capitalization of these successful companies, leading to an appreciation in their 'acquisition currencies'. In turn, this has facilitated merger and acquisition that has allowed the companies to grow and reap

---

[39] See the previous footnote for further details.

economies of scale and scope. Capital markets have also impacted on labour markets in a way that has often benefited successful telecoms operators. One important mechanism has been a rising share price that has allowed the successful operator to provide incentives for new staff to be lured from other companies, often the incumbents. This has alleviated possible human resource constraints on growth. At the same time these processes have brought about a process of consolidation in the industry.

A fifth force has been the increasing intensity in inter-firm competition. In large part, this has resulted from the rapid rate of new entry into the telecoms services industry. Capital markets, as just shown, have facilitated this entry. But equally important have been low technological barriers to entry. These barriers have been lowered by specialist equipment and technology suppliers who have provided the 'black-boxed' technologies that new operators have purchased and integrated—usually with the help of these suppliers—into networks and complementary systems. This explains how companies, with little previous experience in telecoms, have managed to enter the industry. It also explains why they have managed to do so without undertaking much R&D. Examples are WorldCom, established by Bernard Ebbers and colleagues, none of whom had a background in telecoms; Qwest, established by oil-billionaire Philip Anschutz; Teleport and COLT, established by Fidelity, the largest mutual fund in the US; and Mannesmann telecoms, set up by a company previously involved in heavy engineering.[40] The increased intensity of competition, in turn, has driven down prices and increased the rate of introduction of new and improved services.

The sixth and final force operating behind the backs of the telecoms operators comes from the economics of the Infocommunications Industry. Like some other industries, this industry is characterized by high fixed costs coupled with low marginal costs. This has further important consequences. High fixed costs create the necessity to find many customers and generate a large amount of traffic. The more an operator can do this the lower will be its average costs as its fixed costs can be spread over a larger number of customers and amount of traffic. This generates economies of scale. At the same time, if there is sufficient competition, low marginal costs will mean that prices will tend to be low. This is precisely what has been happening with AT&T, BT, and NTT, as was shown earlier. As competition has increased with new entry, their prices have fallen causing revenue and profit problems in their traditional core areas. The new growth areas of mobile, data, and Internet have offered some respite, largely as a result of the explosion of demand in these areas. However, the economics of the Infocommunications Industry suggests that this respite may be temporary.

Further relief may come from economies of scope, where firms benefit from being able to produce jointly a number of products/services more cheaply than if they were produced singly by separate companies. For example, AT&T has

---

[40] The entrepreneurs that established these companies and their entry strategies are examined in Chapters 6 and 7.

Table 3.20. AT&T, BT, and NTT: three questions about the past

Question 1	To what extent is the performance of the three companies due to unavoidable external circumstances, to what extent to mistaken company decisions?
Question 2	To what extent have national institutional factors influenced the performance of the three companies?
Question 3	Are the three companies super-adapters or dinosaurs caught in a climate change?

argued that it can more competitively provide a bundle of services—containing voice, local, long distance, and international services, Internet access, and mobile, together with single billing—than rivals who can only offer some of these services. The problem, however, is that convincing evidence to support this hypothesis is absent, as AT&T was forced to acknowledge when it decided to break up into four independent companies.

It may be concluded, therefore, that these six forces combined have constituted powerful forces with which AT&T, BT, and NTT, as well as all the other telecoms operators, have had to contend.

REFLECTIONS

Questions About the Past

Three key questions may be posed about the past performance of AT&T, BT, and NTT. These questions are shown in Table 3.20.

Question 1. It is clear from the information provided above that the three companies have recently been going through a difficult time. It is often tempting for both analysts and capital markets to lay the blame squarely at the door of a company's leadership. Capital markets are notoriously unforgiving judges. But, before blame is apportioned, it is necessary for a careful analysis that allows us to discriminate between unavoidable external circumstances and mistaken decisions.[41]

One way of conducting such an analysis is to identify the key decisions that the company's leadership took and then formulate hypotheses and examine evidence in the attempt to arrive at a judgement. This method will not be adopted here. Instead, we will take another route, identifying another company that by the common consent of analysts and capital markets has been extremely

---

[41] Indeed the latter should be divided into those decisions where the mistake should have been seen and corrected *ex ante* and those where a mistake that turned out, *ex post*, to be incorrect could not reasonably have been foreseen beforehand.

Table 3.21. WorldCom: performance in major markets

Market	Revenue (1999)	Percentage of total revenue	Projected annual growth rate for next 3 years
Voice	$21.0 bn	43%	3%
Data	$7.5 bn	20%	30%
Internet	$3.5 bn	10%	40%+
International	$1.7 bn	5%	30%+
Share price	October 1995: $13; High: $60; October 2000: $19.		
Percentage of revenue from Internet and data	WorldCom: 30%; AT&T: 16%.		
Market capitalization	WorldCom: $152 bn (July 1999); $52 bn (October 2000) Vodafone: $225 bn (October 2000)		

Source: Bloomberg Financial Markets; company reports.

successful during the last two decades of the twentieth century. This company will be used as a benchmark against which to compare AT&T, BT, and NTT.

The WorldCom benchmark. The benchmark company is WorldCom the vital statistics of which are shown in Table 3.21.

WorldCom has been widely acknowledged as one of the most successful telecoms companies in the world. Launched in 1983 in out-of-the-way Hattiesburg, Mississippi by a handful of entrepreneurs, none of whom had significant prior experience in telecoms, the company that would later be called WorldCom quickly launched itself on a high-growth trajectory. Until the late-1980s the company grew rapidly and highly profitably by re-selling capacity on AT&T's network.

Under Bernard Ebbers, within a context of growing telecoms capacity[42] and entry by numerous other resellers, the company soon realized that the opportunity for arbitrage created by reselling was coming to an end. The decision was accordingly made to transform the company into a network operator. Ebbers discovered that he was also a talented financial operator. Using the company's burgeoning share price as an 'acquisition currency' Ebbers launched into a string of acquisitions that by the close of the century would number around seventy. These included some of the largest companies, notably MCI, the second largest long-distance company in the US after AT&T.

By 2000 WorldCom was the second-largest provider of long-distance voice services in the US with 25 per cent of the market; the largest carrier of data in the US with around 50 per cent of the market; and the largest provider of internet access to corporate customers in the US. According to influential telecoms analysts, such as Jack Grubman of Salomon Smith Barney, by 1999 WorldCom boasted the best global footprint in the world. For example, WorldCom moved

---

[42] At a time before Internet and data demand took off.

Table 3.22. WorldCom: what went wrong?

Falling revenue and profit from core voice services
Falling profit margins in data transport
Time before new growth areas contribute compensating
  revenue and profit
Falling share price
Regulators constraining the 'growth by acquisition'
  trajectory by blocking the acquisition of Sprint
Ebbers' vision failure regarding the importance of mobile

Source: M. Fransman.

rapidly to build a pan-European network and was the second foreign carrier to be awarded a Type 1 licence in Japan, allowing it to compete with NTT.

With such an enviable performance one might have expected, as analysts such as Grubman predicted, that WorldCom would go from strength to strength. However, in the fourth quarter of 2000 WorldCom's advance came to a sudden halt. Revenue growth slowed significantly from double digit figures to around 8 per cent. More importantly, profits, the company announced, were likely to be about half the $1.4 billion that analysts were expecting. WorldCom's share price fell by 20 per cent in a single day, reducing the company's market capitalization by $17 billion.

While in July 1999 WorldCom's market capitalization was $152 billion, making it the fourteenth most valuable company in the world, by October 2000 its value had fallen to $52 billion. In strong contrast, a telecoms company that did not feature in the list of most valuable companies in July 1999, by October 2000, was worth $225 billion, the most valuable in Europe and the seventh most valuable in the world. The company was the mobile 'single play', Vodafone.[43] Ebbers, a man better known for his cowboy hat and swagger than for his humility, admitted in a meeting with analysts on 1 November 2000: 'I've let you as investors down. I've let myself down'.[44] On 30 April 2002 after a profits warning that shocked in Wall Street resulting from reduced internet earnings, Ebbers stepped down as CEO of WorldCom.

What went wrong with WorldCom? Six important reasons are summarized in Table 3.22.

The first thing that went wrong was falling revenue and profit from core voice services. 'There are more than 500 long-distance providers [in the US] today, driving prices down as low as 3 cents a minute for big corporate customers. WorldCom's share of the long-distance market is expected to fall to 15 per cent

---

[43] In November 2000 Vodafone announced that for the first half of the year revenue increased by 32 per cent, pre-tax profits increased by 107 per cent to £1.8 billion. The day of the announcement Vodafone's shares climbed by 10 per cent.

[44] *Business Week*, 20 November 2000, p. 84.

by 2005 from 25 per cent [in 2000].'[45] Second, as more competitors have also entered the data transport market, profit margins have fallen, for precisely the same reason as in the voice market. Third, new high growth, high profit margin businesses are not yet managing to compensate for lower profits from voice and data markets.

Fourth, WorldCom's falling share price is preventing the company from reaping the benefits it previously derived from its rising share price. Specifically, it has caused a depreciation in the company's 'acquisition currency' (although lower share prices all round has made some acquisition targets cheaper) and it has made it more difficult to retain and perhaps motivate staff. Furthermore, it has made WorldCom itself more vulnerable to takeover, once unthinkable. Fifth, not only is acquisition limited by the falling share price, it is also constrained by regulators in the US and Europe who have blocked WorldCom's intended takeover of Sprint, the US's third largest long-distance provider and a significant mobile operator. Sixth, WorldCom's performance has also been negatively affected by the failure of Ebbers to appreciate early enough the importance of mobile communications. In this respect the comparison of WorldCom, started in 1983, and Vodafone, begun in 1985, is instructive. Vodafone's substantial market capitalization advantage compared to World-Com, already referred to, is primarily a reflection of the former's lack, as a single-play mobile company, of slow growing, low profit margin businesses.

Implications for AT&T, BT, and NTT. What are the implications of the WorldCom benchmark for answering Question 1 in Table 3.20? Significantly, the first five causes of WorldCom's problems summarized in Table 3.22 are very similar to four of the five 'sources of pain' suffered by AT&T, BT, and NTT, summarized in Table 3.12. The implication is that it is the same unavoidable external circumstances that have caused significant problems not only for AT&T, BT, and NTT, but also for WorldCom.

But although it is clear that these unavoidable external circumstances have been the cause of significant difficulties for all four companies, it still leaves open the question for each of them regarding the extent to which their own mistaken decisions have contributed to these difficulties. This complex issue will be further analysed within the context of mobile communications.

The case of mobile. It is clear that some mistakes have negatively affected the performance of the companies. One example is BT's handling of its mobile subsidiary, BT Cellnet.[46] The mistake here was not to realize as soon as Vodafone the coming importance of mobile communications, not to give its mobile subsidiary sufficient autonomy from the parent company, and not to understand quickly enough the importance of globalization.

---

[45] *Business Week*, 20 November 2000, p. 84.

[46] Until the late 1990s Cellnet was 60 per cent owned by BT and 40 per cent by Securicor. Thereafter, BT assumed full ownership. In 2000 BT Cellnet was incorporated with BT's other mobile interests in the UK and abroad into the BT Wireless business. In 2001 BT Wireless was completely separated from BT and renamed mmO2.

That Vodafone got these things right, however, may be as much due to the company's specialization and scope as to 'visionary foresight' on the part of Sir Gerald Whent, the man who was responsible for many of Vodafone's key formative decisions in the late 1980s and early 1990s. As one of only two players in the UK's fledgling mobile market from 1985, Vodafone, drawing on the defence background of its parent, Racal, quickly travelled down the learning curve in establishing its analog network. Cash flow and profits soon came flowing in. In deciding how to invest these, Vodafone considered alternatives such as television. However, the company soon came to the conclusion that it would be more profitable to keep on doing what it had already learned to do well, namely build and run mobile networks. With the UK providing its 'university education', Vodafone decided to go on and, with local partners, do the same thing abroad. This set the company, earlier than its counterparts, on the road to globalization that by the end of the century would see it as the best-positioned mobile company in the world.

It is tempting, with hindsight, to attribute Vodafone's success to the 'brilliant vision' of Sir Gerald and his co-decision-makers. But hindsight can be misleading. Another way of looking at the situation is to see Sir Gerald and his colleagues as *constrained* by the resources of their company and these constraints as the main determinants of the strategic decisions that Vodafone made. To put it more concretely, Sir Gerald and his colleagues made the decision to press on with mobile and become a 'single mobile play' for the simple reason that they had proved in the UK that this was what they were good at. Furthermore, the UK experience had also shown that mobile communications could be a profitable line of business. As early comers, they were also well positioned globally, at a time when business interests around the world were beginning to realize that mobile was becoming an attractive business opportunity and when they needed a technically competent partner in order to help them roll out their own networks.

In short, what else could Vodafone have done? The option of starting to build high-speed digital networks could not have appeared very attractive to Sir Gerald. This was already being done by BT and Mercury and a number of new entrants were already entering this area in the liberalized British market. This option would have taken Vodafone too far from the distinctive competence in mobile that it had already, by 1988, begun to accumulate. Besides, it would have been too costly.[47]

Seen in this light, BT's failure to 'see' the coming importance of mobile the way Vodafone did may not be too damning. Furthermore, just as Sir Gerald's 'right' decisions seem to have had structural determinants, so BT's 'wrong' decisions over mobile are also to some extent attributable to structural factors. Most important was the entrapment of Cellnet in the web created by the coexistence of the competing and conflicting interests of the company's different businesses.

---

[47] For a detailed analysis of Vodafone, see M. Fransman, 'Vodafone's Rise as a Global Mobile Operator', (mimeo).

However, it would also be wrong to go too far down the 'structural determinist' road. While acknowledging that these structural determinants did have an important impact on the decisions that were made (and failed to be made), there also remain cognitive causes of the outcome. In short, the cognitive frame or mindset of Ian Vallance and Peter Bonfield, as with that of Bernard Ebbers of WorldCom, precluded until the mid-1990s a high priority being given to mobile communications: with hindsight we do know that this was to cost the companies dearly.

But can these leaders be criticized for not having constructed 'better' cognitive frameworks that incorporated the increasing evidence at this time that mobile was becoming an important high-growth market? The answer to this question must be yes. To some extent it is like the 'IBM Paradox', where IBM, the information-processing company *par excellence*, failed to allow its belief in the importance of mainframe computers to be dented by the growing information available regarding the increasing significance of personal computers.[48]

We conclude, therefore, that BT did make mistakes regarding mobile—the same is certainly true of Bernard Ebbers who only came around to realizing that mobile was important in 1999. AT&T was also rather slow to get into mobile, though much faster than Ebbers. It was only in 1992 that AT&T seriously entered the mobile market with its acquisition of McCaw. NTT entered at the same time as BT. However, partly under regulatory pressure, NTT spun off its mobile subsidiary, NTT DoCoMo, as a semi-independent company. This highly innovative subsidiary soon became NTT's outstanding success story, a success enhanced by the company's outstanding success in mobile Internet access with its i-mode service (as examined in detail in Chapter 9).

Question 2. National institutional factors have also had an influence on the performance of the three companies. This is seen most clearly in the contrasting cases of regulatory influence on AT&T and NTT.

In AT&T's case, regulatory intervention led to the divestiture of the company in 1984. This involved the separation of AT&T's long-distance operations, together with Bell Laboratories, from the regional and local operations that were divided amongst the seven so-called Baby Bells. With this legacy, and with the regulatory restraint preventing AT&T from entering regional and local markets, the company, as noted earlier, concentrated its efforts on internal decentralization and on realizing synergies between the three areas of networks, equipment, and computing. When this strategy failed, the company voluntarily trivested itself in September 1995, forming the new AT&T, Lucent, and the de-merged NCR.

Michael Armstrong took over as CEO of AT&T from Bob Allen in 1997 in the context of a changed regulatory environment, where, with the intention of the

---

[48] The 'IBM Paradox' is analysed in Chapter 1 of M. Fransman, *Visions of Innovation*, Oxford University Press, 1999.

US authorities to encourage competition in local markets, it became acceptable for AT&T to think of getting local access networks. Armstrong's main strategy was to acquire cable TV companies and this he did, investing around $110 billion at the end of the 1990s. This strategy was informed by a cognitive framework emphasizing the importance of synergies. Specifically, AT&T believed that its competitiveness lay in becoming a 'one-stop-shop' offering local, long-distance, and international voice and data services, Internet access, and mobile communications. However, this cognitive framework and the strategy that it supported were wrecked on the rocks of falling profitability and share price.[49] The break up of AT&T into four separate companies, announced in November 2000, signalled an end to this strategy since it would be up to the independent companies to decide how best to play their cards.[50]

It is clear from this brief account of the evolution of AT&T that regulatory intervention has played an important role as one of the main factors shaping the strategy and scope of the company. The same is true of NTT.

A significant proportion of the time and attention of NTT's senior executives has been tied up since the early 1980s in a protracted battle with the Japanese telecoms regulator, the MPT. Furthermore, the absence in Japan of a separation of the policy-making and regulatory functions has meant that only one government body, the Ministry, has responsibility for both functions. Accordingly, there is no counterpart of the regulatory authorities in Japan, like the FCC in the US and Oftel in the UK. One of the consequences is that the regulatory process in Japan has been less transparent than in the US and UK and has been far more politicized and less rule-bound.

During this time, while MPT pushed for the breakup of NTT, the company fought to prevent this from happening.[51] The inevitable result has been a series of politically driven compromises, involving not only the two protagonists, but also other ministries such as the Ministry of Finance (that is still a majority owner of NTT's shares), the Ministry of International Trade and Industry (that is concerned with Japanese competitiveness and international trade), and

---

[49] Significantly, the Baby Bells, and in particular Bell Atlantic (now Verizon) and SBC (now Cingular), have performed much better than AT&T largely because their regional monopoly on local services has held up much longer than AT&T's in long-distance.

[50] Even after breakup, however, Michael Armstrong still hopes that various contracts can be established between the independent companies so as to realize some cross-company synergies.

[51] One of the consequences of the idiosyncratic regulatory regime in Japan and its highly politicized nature is that the motives of regulation have become obscured. While the normal regulatory concerns of dealing with a dominant player and establishing a level playing field in the competitive process are apparent in MPT's interventions, there appears to be more to the Japanese regulatory story than just this. Some political scientists have seen in the conflict between MPT and NTT the attempt of part of the Japanese bureaucracy to expand its sphere of influence. Whether or not this is true, it is clear that the absence of an arm's-length regulatory authority such as the FCC in the US and Oftel in the UK has meant that the Japanese regulatory process lacks the transparency of its Western counterparts.

political parties, particularly the Liberal Democratic Party, that have some influence over the bureaucracy.

There have been two undesirable consequences of this politically tortured process. The first is that crucial decisions about the structure and scope of NTT have been driven as much, if not more, by political necessity and compromise as by strategically-driven business logic. This may very well explain the fact, noted in Table 3.4, that NTT's divisional structure is based far more on geographical considerations than AT&T's or BT's, which are more concerned with market structure. The second undesirable consequence has been that NTT's leaders have been forced to devote more time and give more of their limited attention than they would have liked to regulatory issues.

By contrast, BT has come off relatively unscathed. During the first part of the liberalization process, after the company was privatized in 1984, BT was put in the enviable position of being part of a duopoly with the Cable & Wireless subsidiary, Mercury. Similarly, from 1985 to 1993 BT's mobile subsidiary, Cellnet, and Vodafone were granted duopoly status. BT was never broken up, or seriously threatened with breakup, by organs of the British government. It is true that BT's prices were regulated with the requirement that the company reduce prices more than in line with inflation. But as an embarrassing dispute over local loop unbundling with the European Commission revealed in 2000, there are grounds for believing that BT has been more 'lightly' regulated than its major European counterparts. In BT's case, therefore, it seems reasonable to conclude that the company's difficulties are more the result of the 'forces operating behind the backs' of the company, analysed earlier, and the consequence of the company's own structure and strategy, than the effect of adverse regulatory decisions.

We conclude, therefore, in answer to Question 2, that national institutional factors, particularly regulation, have been important for the three companies and have influenced their performance.

Question 3. We are now in a position to tackle the final, daunting, question: Are the three companies super-adapters or dinosaurs caught in a climate change?

Let us begin with the dinosaurs. The obvious problem with the dinosaur metaphor is that dinosaurs, like other biological organisms, are limited by their genetic make-up in adapting over the short term to environmental changes. The same is obviously not the case with our three 'organisms', AT&T, BT, and NTT. As this chapter has shown in great detail, all three companies have made significant attempts to adapt to their changing environment. They have diversified into new areas of business—such as data, the Internet, and mobile—and they have de-emphasized previously core activities such as plain old voice. They have rapidly adopted, and in some cases developed, new technologies and forms of organization. And they have significantly changed their internal corporate structures moving, as shown in Table 3.4, from hierarchical integration to decentralization plus empowerment and, in AT&T's, and to some extent BT's, case, to breakup.

It is clear, therefore, that all three companies have been adapters. But have they been super-adapters? The answer, of course, depends on what is connoted by the prefix 'super'. Super-adapters implies that they have been able, not only to adapt, but also to adapt successfully so that they are able both to survive and to prosper.

One of the important lessons to be learned from the analysis in this chapter of the dynamics of the telecoms industry through the prism of three of the most important incumbents is that even though they have adapted to their changing environment they have to a significant extent been caught out not only by their 'genes' (i.e. their internal organizational structures and practices) but also by their past. More specifically, all three companies have not only a legacy of older technologies but also a legacy of markets and revenues. All three are still, by the turn of the millennium, dependent on revenue and profit from plain old fixed voice telephony. A rapid increase in intense competition—the result of significant new entry fuelled by vertical specialization in the industry—has quickly pushed down prices and profitability in this traditional area. Unfortunately, the new growth areas of data, the Internet, and mobile, while adding significantly to revenue growth and profitability, have been insufficient in the short term to compensate for the damage done to traditional voice. Companies that more through good fortune than superior vision-construction managed to avoid the 'plain old voice dilemma' have been the winners in these shifting sands. Vodafone is perhaps the best example.

But the fact that the plain old voice dilemma has hit not only AT&T, BT, and NTT, but also the relatively new entrant, WorldCom, suggests that there are limits to the effectiveness of adaptation in this case. It may be concluded, therefore, that although they may have done a 'super' job in adapting, they are still left with significant challenges ahead in prospering in the new telecoms world of the new millennium.

CONCLUSION

The simultaneous advent of liberalization, though for different reasons, in Japan, the UK, and the US in the mid-1980s raised a crucial question. Would the incumbents of these three countries—NTT, BT, and AT&T—survive? Although all three companies were highly innovative, as is shown by the technological advances that they pioneered, many of which still drive the Infocommunications Industry, they were accorded a special position by their governments. Specifically, they were publicly regulated monopolies. While this did not insulate them from political pressures to improve telephone services, and although they competed to be the first to make and introduce new generations of technology, it did mean that they were protected from competition in service, equipment, capital, and labour markets. Many wondered whether they would be able to survive competition from the new generation of new entrants that would be unleashed by liberalization of markets and the forces that lowered entry barriers into these markets.

The present chapter may be taken as a summary, at the beginning of the new millennium—some fifteen years after telecoms liberalization was pioneered in Japan, the UK, and the US—of the state of play in the Telecoms Industry. What may be concluded about the viability and performance of these incumbents?

The first observation is that although two of them—namely, AT&T and BT—received something of a battering in the first year of the new millennium with NTT also suffering, all the companies are still here. Moreover, they are strong contenders in many of the new growth segments of the Telecoms Industry. NTT's outstanding success in mobile communications through its subsidiary, NTT DoCoMo, is a notable achievement and it is hard to find similarly outstanding accomplishments on the part of the other two incumbents outside market segments where they still enjoy, like some of the former Baby Bells in the US, de facto monopoly status. However, as Tables 3.13 and 3.14 show, AT&T and BT are showing healthy growth in the promising new market segments. Although many questions remain about the impact of their new organizational structures, all three companies are still in with a fighting chance as the new millennium dawns.

If there is one area, however, where this optimistic conclusion might be qualified it is in the area of globalization. This chapter has shown that AT&T and BT have not been very successful in globalizing. And NTT has not been allowed to operate globally until July 1999. However, even WorldCom, a more successful globalizer than the three incumbents, only has 5 per cent of its revenue coming from outside the US. In this area, Vodafone is the outlier although the strength it derives from its global operations illustrates the potential contribution of globalization as a source of corporate growth.

The use of WorldCom as a mirror in which to view the performance of AT&T, BT, and NTT has been illuminating. By using as a benchmark the company that is, globally, one of the most successful of all the new entrants the perhaps surprising result has emerged that to a large extent the three incumbents have been victims of circumstance rather than of their own mistaken decisions. As was shown clearly, WorldCom has been afflicted by essentially the same causes as those that have negatively impacted on the three incumbents, causes that have operated 'behind the backs' of all the telecoms operators in the Telecoms Industry. However, it has been insisted in this chapter that the cognitive framework, or 'vision', of the leaders of the companies has also played an important role. Also, in some instances—as with IBM's mistaken vision of the might of the mainframe—their visions have been incorrect and have needed revision.

Will these incumbents survive and prosper through the next quarter of a century? On this question the jury must still be out, although the start of the century sees them in with a fighting chance.

4

Deutsche Telekom: Europe's Giant Struggling to Globalize

In Chapters 1 and 2 the main forces driving the Telecoms Industry in the Internet Age were examined. The fortunes of the three main incumbent network operators who were the first to face competition in their domestic markets from around the mid-1980s—AT&T, BT, and NTT in the US, UK, and Japan respectively—were analysed in Chapter 3. The UK aside, the other countries in the European Union were far slower to liberalize their telecoms markets and introduce competition. Indeed, it was only on 1 January 1998 that competition was officially introduced by the European Union.

How have the main incumbent network operators from the latter European countries adapted to the changing circumstances of the Telecoms Industry in the Internet Age? In this chapter a start is made in answering this question by examining in detail how Deutsche Telekom has coped. In the following chapter the same is done for France Telecom.

INTRODUCTION

Deutsche Telekom is Europe's largest incumbent telecoms network operator and service provider from Europe's largest economy. However, size alone has not been a guarantor of performance. Indeed, by February 2001 Deutsche Telekom had lost 70 per cent of its stock market value in a year, making the company the worst performer in the German blue-chip DAX index. Deutsche Telekom's stock market performance has been significantly below that of its former ally and now bitter rival, France Telecom (FT). But Deutsche Telekom is pressing on with its ambitious attempt to transform itself into one of the few genuinely global

I wish to acknowledge the help I have received in writing this chapter from Klaus Rathe's work on Deutsche Telekom.

telecoms network operators and service providers, an ambition that has already been abandoned by the company's European counterparts, BT and FT. Is Deutsche Telekom likely to succeed in the tough task it has set itself? What are the company's strengths and weaknesses? These are the questions that will be analysed in this chapter.

PERFORMANCE OF DEUTSCHE TELEKOM: AN OVERVIEW

The stock market provides one possible measure of a company's performance. In theory, (under a restrictive set of assumptions) stock markets take all available information into account and on the basis of this information determine the price of a company's shares that is equal to the discounted value of that company's future earnings. In practice, of course, things are not nearly so simple. The 'given information' set is not homogeneous or complete but is usually contradictory and incomplete. Interpretive ambiguity rules. Financial analysts usually disagree regarding 'the essentials' for any company. Stock markets are subject to 'fashionable sentiments' and national circumstances often distort comparisons between companies headquartered in different countries.

However, while all this is true, this does not mean that stock market measures are valueless in seeking a measure of performance. While it may be impossible at any point in time to measure accurately the 'true value' of a company through its market capitalization, over longer periods of time there is likely to be a fairly strong correlation between a company's earnings growth and its growth in share price. The experience over the 2000/2001 period of the so-called 'dotcom' companies, however, shows how treacherous such relationships are over shorter periods of time.

How well has Deutsche Telekom's share price performed relative to other telecoms companies? Table 4.1 provides some data in answer to this question.

Several interesting points emerge from Table 4.1. The first is the remarkable fact that, with only two exceptions, the telecoms companies shown in the table experienced an almost identical fall in their share price over the one year period ending on 24 February 2001. For seven of the nine companies their share price fell around 55 per cent—from 53.47 per cent for Vodafone to 59.39 per cent for AT&T. This almost identical fall occurred despite very different earnings per share and price/earnings ratios, as shown in the last two columns of Table 4.1. In addition, the companies are drawn from very different parts of the Telecoms Industry, the first five—referred to in Table 4.1 as the Big Five Incumbents—are network operators and service providers, while the third group are telecoms equipment suppliers. Vodafone is to be distinguished from the Big Five Incumbents since it is a 'pure mobile play'.

The two exceptions are Deutsche Telekom and Lucent. While over the one-year period Deutsche Telekom's share price fell by 68.90 per cent, Lucent's fell

Table 4.1. Deutsche Telekom: relative share price performance

Company	1-year change in share price (to 24/02/01) (%)	5-year change in share price (to 24/02/01) (%)	EPS (Mar. 2000)	*P/E*
Big 5 incumbents				
AT&T	−59.39	−29.64	0.53	12.60
BT	−54.99	+58.19	1.14	29.30
Deutsche Telekom	−68.90	+6.05	0.01	N/A
France Telecom	−54.67	+76.17	N/A	29.90
NTT	−56.62	−14.74	N/A	49.2
Mobile leader				
Vodafone	−53.47	+278.62	N/A	55.5
Equipment suppliers				
Cisco	−55.12	+454.96	0.14	39.70
Lucent	−76.22	+73.11	0.25	−22.30
Nortel	−57.29	+254.49	0.12	26.20
Nasdaq	−44.16	+111.56		

Notes: EPS = earnings per share; *P/E* = price earnings ratio.

Source: http://quote.bloomberg.com 24/02/01.

even more by 76.22 per cent. Deutsche Telekom's share price, therefore, fell significantly more than all the other telecoms network operators and service providers, including Vodafone.

The second point to emerge from Table 4.1 is that the companies exhibited remarkably different share price changes over the five-year period to 24 February 2001. The two leaders in the group of nine companies as a whole were Cisco, far and away the best performer, showing a 454.96 per cent gain and Nortel recording a 254.49 per cent improvement. Of the network operators and service providers it was only Vodafone—the pure mobile play—that achieved a comparable increase in share price, namely 278.62 per cent.

The five-year share price performance of the Big Five was decidedly mixed. The best performer was France Telecom that registered a 76.17 per cent increase, followed by BT with 58.19 per cent. Deutsche Telekom was by far the worst performer on this measure of the European incumbents, with only a 6.05 per cent increase. Of the non-Europeans, NTT recorded a 14.74 per cent fall, a result that is significantly influenced by the stubborn recession of the Japanese economy throughout the 1990s and into the new millennium. AT&T was by far the worst performer of the Big Five, recording a 29.64 per cent decline over the five-year period. The reason for AT&T's performance was its greater degree of dependence on long-distance voice telephony compared to its other four counterparts in the Big Five. Of all the five it was only AT&T that was deprived of an initially dominant position in local telecoms access when, in the divestiture of the company in 1984, AT&T was separated from the seven Bell regional holding companies, the so-called Baby Bells.

The third point to emerge from Table 4.1 is that over the one-year period all the telecoms companies—network operators as well as equipment makers—performed worse than the NASDAQ index of US-quoted 'technology' shares.

With share price as an indicator of performance,[1] therefore, Deutsche Telekom has performed significantly worse than almost all of the other telecoms companies included in the table. In the rest of this chapter the reasons for this inferior performance will be examined.

HISTORICAL BACKGROUND

The company Deutsche Telekom AG was established in Bonn on 2 January 1995. This followed a long period, beginning in the early 1980s, when the liberalization of telecoms was put on the German political agenda. In 1987 telecoms liberalization began to be seriously discussed at the European level with the publication of a European Union green paper making explicit the intention to establish a common market in Europe for telecoms services and equipment. In August 1996 the new German Telecommunications Act came into force, ending Deutsche Telekom's monopoly on networks with immediate effect. This was followed later on in the same year by the initial public offering of 26 per cent of the company's shares. On 1 January 1998 in line with the European Union directive establishing liberalized telecoms markets throughout the Union, Deutsche Telekom's monopoly on fixed-network voice telephony services was ended. On the same day a new regulatory authority was established, The Regulatory Authority for Telecommunications and Posts (RegTP) under the authority of, but operationally separated from, the Federal Ministry of Economics (BMW). On 27 June 1999 Deutsche Telekom had its second public share offering.

The Role of the German Government in Deutsche Telekom

In Table 4.2, the holding of the German government in Deutsche Telekom is shown.

As shown in Table 4.2, as of July 2000 the German government held 60 per cent of the shares of Deutsche Telekom. This included 43 per cent of the shares held directly by the Federal Government of Germany and a further 17 per cent held by a federal financial institution, Kreditanstalt fur Wiederaufbau (KfW). The 60 per cent held by the government in July 2000 was down from

---

[1] Other indicators are used later in analysing Deutsche Telekom's performance.

Table 4.2. Deutsche Telekom: share ownership structure

Date	Shareholder (percentage holding in brackets)
December 1999 (%)	German government—43
	Kreditanstalt fur Wiederaufbau (KfW)—22
	France Telecom—2
	Institutional investors—18
	Private investors—15
July 2000 (%)	German government—43
	KfW—17
	Institutional and private investors—40

65 per cent held at December 1999. It is the declared intention of the government to significantly decrease its holding in Deutsche Telekom, an issue that emerged as a complication with the US government when Deutsche Telekom made a bid for the American seventh-largest mobile company, VoiceStream.

The large stake that the German federal government has in Deutsche Telekom, coupled with the fact that some 2.8 million Germans hold shares in the company, has meant that the company's fortunes have become something of a political football in Germany. For example, at the beginning of 2001, at a time when Deutsche Telekom's share price had fallen by more than 60 per cent and the company was struggling under a debt burden of $51 billion, the opposition Christian Democratic and Free Democratic parties demanded the resignation of Ron Sommer, Deutsche Telekom's CEO. However, the German chancellor, Gerhard Schröder, rushed to Sommer's defence arguing that Deutsche Telekom's share price was undervalued and congratulating Mr Sommer in 'setting up the company well both domestically and abroad'.

Regulation of Deutsche Telekom

The regulation of Deutsche Telekom has always been a politically complex affair. Under the conservative government of Chancellor Kohl the company tended to be seen as bureaucratic and inefficient and as an ideal candidate for attempts to shrink the state sector and expand the private sector. This political perspective was reflected in relatively harsh regulation imposed on Deutsche Telekom by the new regulator, RegTP, according to some observers one of the harshest regulatory regimes in the major industrialized countries. Regulation, however, continued to be a political football and attitudes changed sharply with the accession to power of the centre-left government of Gerhard Schröder which saw Deutsche Telekom as an important champion of German globalization.

DEUTSCHE TELEKOM'S FOUR PILLARS

In 1999 Deutsche Telekom under Ron Sommer made the decision to reorganize the company into four 'pillars' in order to increase strategic and market focus. These four pillars are shown in Table 4.3.

As is clear from Table 4.3, Deutsche Telekom is organized into four businesses: mobile, Internet, system solutions, and fixed network. Each of these pillars is analysed in more detail later in this chapter where both their strengths and their weaknesses are examined. An overview of the company's revenue distribution among these pillars is given in Table 4.4.

Table 4.4 reveals two important facts about Deutsche Telekom. The first is the continuing quantitative importance of T-Com, the company's fixed network pillar. In 2000 this pillar contributed two-thirds of Deutsche Telekom's total revenue. However, the importance of revenue from this source is decreasing rapidly. Only a year earlier, revenue from T-Com accounted for no less than 78 per cent of Deutsche Telekom's total revenue.

The second fact about Deutsche Telekom is the increasing significance of the contribution to total revenue of T-Mobile. In 2000 this pillar contributed almost a quarter of Deutsche Telekom's total revenue, up from only 13 per cent a year earlier. While Table 4.4 provides an overview, an even more detailed account of the breakdown of Deutsche Telekom's revenue (at a more disaggregated level than the pillars) is given in Table 4.5.

Further important characteristics of Deutsche Telekom emerge from Table 4.5. To begin with, a disaggregation is provided of some of the most important components of the T-Com pillar. More specifically, network communications was, in 2000, the most important single source of revenue, accounting for just over a third—38.3 per cent—of total revenue. However, revenue from network

Table 4.3. Deutsche Telekom's four strategic pillars

T-Mobile	T-Online	T-Systems	T-Com
Mobile telecoms	Internet services	Data/IP and system solutions	Fixed network, telecoms infrastructure, CATV, access

Table 4.4. Deutsche Telekom: distribution of revenue by pillar

Revenue (% of total)	T-Mobile (%)	T-Online (%)	T-Systems (%)	T-Com and others* (%)
Q3 2000	22	2	9	67
Q3 1999	13	1	8	78

*Others include carrier services, value added service, broadcasting and broadband cable, terminal equipment, MATAV, and SIRIS.

Table 4.5. Deutsche Telekom: growth in revenue by area

Area	Q1–3, 2000 (millions of Euros, % of total revenue in brackets)	Q1–3, 1999 (millions of Euros)	Change in %
Network communications	11,183 (38.3)	12,522	−10.7
Carrier services	2,959 (10.1)	1,972	50.1
Data communications	2,573 (8.8)	2,123	21.2
Mobile communications	6,421 (22.0)	3,292	95.0
Broadcasting and broadband cable	1,432 (4.9)	1,384	3.5
Terminal equipment	762 (2.6)	917	−16.9
Value-added services	1,363 (4.7)	1,405	−3.0
International	1,537 (5.3)	1,157	32.8
Other services	992 (3.4)	804	23.4
Total	29,222		14.3

Source: Deutsche Telekom, Group Report, 1 January to 30 September 2000, p. 8.

communications fell by 10.7 per cent compared to one year earlier. This high-lights the dramatic fall in Deutsche Telekom's core business, an issue that is taken up again later.

Mobile communications, as already shown in Table 4.4, accounted for 22 per cent of total revenue, making this the second most important source of revenue in Table 4.5. However, revenue from mobile communications increased by no less than 95.0 per cent over the previous year, making this by far the most rapidly growing source of revenue.

The second most rapidly growing source of revenue was carrier services, up by 50.1 per cent on the previous year, although this source accounted for only 10.1 per cent of Deutsche Telekom's total revenue. Carrier services were the third most important source of revenue of the areas shown in Table 4.5.

Data communications was the fourth most important revenue source, accounting for 8.8 per cent of the company's total revenue. Despite its relatively small proportional contribution, data communications revenue grew at a respectable rate of 21.2 per cent over the year to September 2000.

International communications was the fifth most significant source of revenue, contributing 5.3 per cent of Deutsche Telekom's total revenue. Revenue from this source was the third most important in terms of growth from September 1999 to September 2000, growing at a rate of 32.8 per cent. This illustrates the relevance for Deutsche Telekom of growing globalization, one of the company's major priorities.

Significantly, broadcasting and broadband cable accounted for only 4.9 per cent of Deutsche Telekom's total revenue and grew at a rate of only 3.5 per cent over the year. This is largely a reflection of the regulator-imposed restrictions on the company's use of its cable TV networks for the purpose of providing broadband

Table 4.6. Deutsche Telekom: falling revenue from long distance

Date	Total revenue (billion Euros)	Long-distance revenue (billion Euros)	Long-distance revenue % total revenue
Q1–Q3 1998	26.0	6.4	24.7
Q1–Q3 1999	25.6	3.4	13.1
Q1–Q3 2000	29.2	2.3	8.0

An international call from Bonn to New York costs the same as a long-distance call from Bonn to Hamburg.

Internet access. Fearing that Deutsche Telekom would wield excessive market power if it were allowed to retain both its copper local loop and its cable TV networks, the German regulator ruled that the company should sell a majority part of its interests in the latter. Accordingly, Deutsche Telekom is now relying on its DSL local networks for its competitive strength in the broadband access market. This important point is considered again later in this chapter.

Finally, the insignificance of terminal equipment as both a contributor of revenue and a source of revenue growth is worth noting. In 2000 equipment accounted for only 2.6 per cent of Deutsche Telekom's revenue although revenue from this source fell by 16.9 per cent during the year to September 2000. This is a reflection of the increasing vertical integration in the Telecoms Industry as a whole, a process that has seen specialist equipment suppliers— such as Siemens in the German case—assume control of the equipment layer of the industry.

Further details are provided in Table 4.6 regarding the decline in revenue from network communications. This table examines Deutsche Telekom's falling revenue from long-distance calls.

Table 4.6 highlights in dramatic terms the decline in one of Deutsche Telekom's core business areas. While long-distance revenue provided no less than one quarter of the company's total revenue in 1998, by 2000 this had fallen to 8.0 per cent. In absolute terms revenue from long-distance calls declined from 6.4 billion Euros to 2.3 billion over this period. This is due to the sharp increase in new technology-fuelled competition that has hit all telecoms service providers, and all the service providing incumbents, in the world. The effects of this competition is dramatically illustrated by the fact that a long-distance call from Bonn to Hamburg now costs the same as a call from Bonn to New York.[2]

CORPORATE ORGANIZATION FROM 2001

The essential characteristics of Deutsche Telekom's corporate organization are described in Fig. 4.1.

---

[2] The forces underlying these trends were analysed in Chapters 1 and 2.

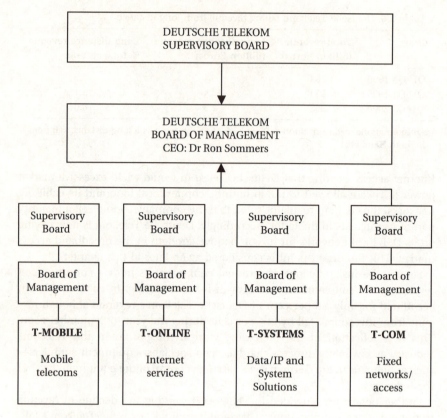

Fig. 4.1. Deutsche Telekom: organizational structure.

Source: M. Fransman.

The following are the main characteristics of the corporate organization of Deutsche Telekom. Taking a 'bottom up' approach, the company is based, in the words of Sommer, on 'four pillars' that together define the company's scope, competencies, assets, and specialization. These four pillars, as shown in Fig. 4.1, are mobile, Internet, data and system solutions, and networks and access.

The declared intention of Ron Sommer and Deutsche Telekom is for each of these four pillars to operate as semi-independent businesses. An attempt has been made to put this declared intention into organizational practice by giving each pillar its own Board of Management (each under its own CEO) and its own Supervisory Board. As we shall shortly see, however, while the intentions and aims behind this organizational design seem relatively clear-cut, in practice Ron Sommer has run into serious difficulties trying to implement this design and run an efficient company.

Deutsche Telekom's Supervisory Board

Further elaboration is required on the role of the supervisory boards, both at the level of Deutsche Telekom as a whole as well as at the level of the constituent pillars. As indicated by the adjective 'supervisory' in the title of the Supervisory Board and by the downward arrow in Fig. 4.1 going from the Supervisory Board to the Board of Management, the role of the Supervisory Board is to oversee the activities of the Board of Management. Under the German system of corporate governance that evolved in the aftermath of the Second World War, for most German companies the Supervisory Board would have strong representation both from the workers in the company concerned as well as from the German banks that provided a significant source of financing for the company's activities. The role of German banks in the Supervisory Board has been compared to the similar role played in practice by Japanese banks, traditionally also an important source of financing for Japanese companies.

Deutsche Telekom's Supervisory Board, however, is different from that of most other German companies in one significant respect. As we have seen in Table 4.2, the German government, rather than German banks, is the most significant single shareholder in Deutsche Telekom. This ownership position is reflected in the relatively large number of German government-related appointees on Deutsche Telekom's Supervisory Board and the correspondingly small number of German bank representatives. Table 4.7 puts this in quantitative terms.

As is evident from Table 4.7, twelve out of the twenty-six members of the Deutsche Telekom Supervisory Board are drawn from the ranks of worker representatives from within the company or from the German trade union movement. Both the vice-chairs of the Supervisory Board were workers' representatives while the chairperson was a member of the shareholders' committee of a German company (Henkel). Seven out of the twenty-six members, including the chairperson, were from the German business community.

Table 4.7. Deutsche Telekom's Supervisory Board, 1999–2000

Interests represented	Number
Workers from Deutsche Telekom	12
German business	7
German government	5
Other	2
Total	26

Source: Calculated by the author based on information provided in Deutsche Telekom, 1999 Annual Report, p. 10.

Significantly, however, only one of these was from a German bank (in fact was the Chairperson of the Managing Board of Dresdner Bank). Five of the twenty-six were from the German government. Of these, three were currently members of the German Bundestag (parliament), two were retired federal government ministers, one was a current federal government minister (state secretary from the Federal Ministry of Finance), and one represented a federal government financial institution that was also a major shareholder in Deutsche Telekom (i.e. Kreditanstalt fur Wiederaufbau (KfW)—see Table 4.2). Two of the twenty-six represented other interests, one being a professor from St. Gallen University and the other from a broadcasting organization. According to Deutsche Telekom's 1999 Annual Report, 'Deliberations [of the Supervisory Board during the year] focused on the [company's] internationalization strategy, making use of opportunities in the dynamic market of new communication technologies, securing competitiveness under the conditions of an asymmetrical regulatory regime and, not least, the consequences of these issues for our staff' (p. 8).

Although important questions arise in this system of corporate governance regarding the strength of the supervisory 'teeth' of the Supervisory Board, it is clear that as the major single shareholder the German government has an important degree of influence. This is so even though Chancellor Gerhard Schröder—in defending Ron Sommer in early 2001 against calls from the two main opposition parties that he be forced to stand down as a result of the company's poor stock market performance—insisted that according to the law Deutsche Telekom was a private company that had to meet market criteria. 'All these opposition politicians who turn to the federal government either have no idea of the law or want to bring Deutsche Telekom closer to the state, which actually harms its development', he declared.[3]

Deutsche Telekom's Board of Management

Table 4.8 shows the functions of the eight-person Board of Management in 1999/2000.

The Plans of Mice and Men Gang aft Astray

It is never possible in an organizational chart to capture the organizational and interpersonal dynamics that lie at the heart of every organization. This is why such charts are always misleading. Unidirectional arrows and diagrammatic hierarchies, purporting to depict power through 'higher' and 'lower'

---

[3] *Financial Times*, 27 February 2001.

Table 4.8. Functions of Deutsche Telekom's Board of Management

Position	Function
Chairman (i.e. Ron Sommer)	Group strategy, communication, auditing, organization
Board member	Sales and service
Board member	International
Board member	Product marketing
Board member	Networks, purchasing, and environmental protection
Board member	Technology and services
Board member	Human resources and legal affairs
Board member	Finance and controlling

Source: Deutsche Telekom, Annual Report, p. 7, valid on March 2000.

juxtapositions, are travesties of the truth rather than accurate portrayals of the truth. Usually, in real organizations, the organizational and interpersonal dynamics, determinants of the fitness of an organization to survive and prosper in its selection environment, are hidden like the submerged majority of an iceberg. The nature of the submerged bulk is itself ambiguous and competing and contradictory accounts usually coexist regarding who actually wields power, propounded by those on the inside of the organization who are allegedly 'in the know'.

Occasionally, however, some of the waters surrounding the iceberg momentarily part, revealing a passing glimpse of what is happening underneath. These events are usually occasioned by internal conflict, when some or all of the warring parties 'go public' in fighting their battles or licking their wounds. Even then, what 'actually happened' is usually still surrounded by a degree of interpretive ambiguity, the account depending on who is giving it. And then, as suddenly as they parted, the surrounding waters close once more, obscuring again the submerged bulk. But at least there was a glimpse, however brief.

One such event was the battle that took place between Ron Sommer, CEO and Chair of the Board of Management of Deutsche Telekom, and the entire top management of one of the four pillars of the company, T-Online. The background of Sommer is given in Table 4.9, although it is not particularly illuminating as an explanation of the conflict that occurred.

A succinct outline of the conflict is provided in Table 4.10.

What is the significance of what may be called the T-Online Affair? In organizational (rather than personal) terms the T-Online Affair illustrates the complex dynamics and contradictions that are usually involved in the processes of decentralization and empowerment in large companies. The complexity of these companies almost inevitably necessitates a process of decomposition so that tasks such as leadership, coordination, and monitoring can be distributed

Table 4.9. Dr Ron Sommer: background of the CEO of Deutsche Telekom

Dr Ron Sommer was born in 1949, in Haifa, Israel. He spent his childhood in Austria and studied mathematics in Vienna, receiving his doctorate in 1971. Dr Sommer began his professional career with the Nixdorf group, first in New York and then in Paderborn and Paris. In 1980 Dr Sommer joined the Sony electronics group as a managing director of its German subsidiary. In 1986 he became Chairman of the Board of Management of Sony Deutschland. In 1990 Dr Sommer became President and Chief Operating Officer of Sony Europa in the same function. In 1995 the Supervisory Board of Deutsche Telekom AG appointed Dr Sommer Chairman of the Board of Management.

Table 4.10. Deutsche Telekom: the politics of corporate decentralization and empowerment[4]

T-Online is Deutsche Telekom's Internet business, the largest in Europe, and one of the company's four strategic 'pillars'. The company was formed from a former on-line service, Bildschirmtext (Btx), originally established in 1983 by the Deutsche Bundespost (Deutsche Telekom's predecessor). In 1995 Btx was turned into an Internet service business with the addition of e-mail and Internet access services. Wolfgang Keuntje was CEO and Chairman of T-Online's Board of Management. Keuntje had worked for Deutsche Telekom since 1994 and had been T-Online's CEO since 1996. On 17 April 2000 9.32 per cent of T-Online's shares were floated. However, serious tensions soon emerged between Keuntje and Ron Sommer, CEO of Deutsche Telekom. At issue was where decision-making power lay regarding T-Online's strategy and acquisitions. According to reports, Keuntje was close to closing a deal to acquire Freeserve, the largest independent Internet service provider in Britain. Sommer, however, intervened and cancelled the deal on the grounds that the agreed price was too high and strategically unsound. This led to severe strains between Sommer and the senior management of T-Online. On 25 August 2000 Sommer fired Keuntje and in the following three months also fired the rest of the top management of T-Online, bar one individual. This was Eric Danke, a Sommer supporter who was the leading figure in the development of the old Btx online service by the former Deutsche Bundespost. For four months T-Online had no CEO. Then, on 1 January 2001 Thomas Holtrop was appointed as the new CEO of T-Online, reportedly well aware that real power would lie with Ron Sommer.

effectively throughout the organization rather than overwhelming those at the apex. However, the process is contradictory. On the one hand, it is necessary in order to alleviate pressure and limited attention span at the apex; on the other hand, it removes some of the effective power that is nominally vested in the

---

[4] This account draws on work by Klaus Rathe.

CEO who, after all, does bear responsibility for the entire organization. It is in dealing with this contradiction that the personality of the CEO does matter, at least in the short run. In Sommer's case the decision was made to impose control from the top, in so far as key issues such as strategic direction, acquisition, and funding were concerned. Other CEOs may have handled the situation somewhat differently. Each CEO, however, has no option but to negotiate this contradiction in the way she or he thinks best for there is no way to avoid it.

DEUTSCHE TELEKOM'S STRATEGY

Deutsche Telekom's overall corporate strategy is shown in Table 4.11.

Several important points emerge from Table 4.11 regarding Deutsche Telekom's overall strategy. The first is that the company is committed to remaining an 'integrated' corporation. This means that unlike AT&T, which under Michael Armstrong made the decision in October 2000 to break the company up into four separate companies, Deutsche Telekom will continue to 'integrate' the activities of its four 'pillar' constituents. In this sense Deutsche Telekom at the beginning of 2001 was in the same camp as France Telecom (FT) and NTT that had also retained, despite important decentralizing reorganizations, an integrated form of corporate organization.

The second point emerging from Table 4.11 is that Deutsche Telekom has committed itself to becoming a global provider. This sharply separates Deutsche Telekom from its two main European counterparts—BT and FT. Both the latter companies, under the pressure of burgeoning debts occasioned by expensive acquisitions and third-generation mobile licences and infrastructure, made the decision at the end of 2000 to focus their attention on Europe while providing global network services to their European large company customers. Despite Deutsche Telekom's similarly heavy debt burden—that by February 2001 had reached $50.8 billion—the company made the bold decision to press on undaunted with its globalization ambitions. This was most clearly evident in the company's ambitious attempt to acquire a majority share in VoiceStream,

Table 4.11. Deutsche Telekom: corporate strategy 1: overall strategy

---

Position DT as an 'integrated, global telematics provider'
Consolidate DT's market leadership in Europe
Focus DT's activities on four business pillars: T-Mobile,
 T-Online, T-Systems, and T-Com
Prioritize strong operational performance in growth areas
Enter key economies to expand DT's global reach
Majority control of mobile assets

Source: Annual Report, 1999; presentations to analysts, April 2000, January 2001, and author's interviews.

the seventh-largest US mobile network operator (discussed in more detail below in the section on T-Mobile).

Third, Deutsche Telekom describes itself as a 'telematics' rather than a telecoms services provider. This description draws attention to the company's fourth and latest 'pillar' to be added to its corporate portfolio, namely T-Systems. The business of T-Systems is to provide integrated information and communications system solutions that cover the converged fields of computing and telecommunications. (As we shall see later, a major German acquisition was chosen as the vehicle for the attainment of this plank of the company's strategy, a decision that has not been without its complications.)

Fourth, Deutsche Telekom has declared its strategic intention to 'consolidate [its] market leadership in Europe'. While this alleged 'leadership' is based more on absolute subscriber numbers in Germany than leading market positions in the other European countries, Deutsche Telekom's strategic intent to become a leading European player is clear. This emerges clearly from the company's recent acquisitions in the rest of Europe, examined later. The fifth point to emerge from Table 4.11, already considered earlier in this chapter, is Deutsche Telekom's organizational focus on its four 'pillar' businesses.

Finally, Table 4.11 also states Deutsche Telekom's strategic decision to immediately acquire majority, rather than minority, control of mobile assets. This sets the company apart from incumbents such as BT, NTT DoCoMo, and Vodafone, that in some cases made the decision to begin with minority holdings and then, where the business proved viable, pursued majority control. The costs and benefits of these alternative strategies, however, will not be further pursued here.

In Table 4.12 Deutsche Telekom's immediate strategic priorities for 2001–02, following from its overall corporate strategy outlined in Table 4.11, are shown.

Table 4.12 expresses Deutsche Telekom's strategic priorities for 2001 and 2002. Clearly, the company attaches high priority to its acquisition of VoiceStream, the seventh-largest mobile operator in the US and, like Deutsche Telekom itself, an adopter of the European GSM (global system for mobile communications) digital mobile standard technology. As will be shown later in

Table 4.12. Deutsche Telekom: corporate strategy 2: strategic priorities, 2001–02

Area	Priority
Acquisitions	Acquiring VoiceStream number one priority
Mobile	Significant push in mobile data: GPRS and T-Motion
Internet	Leveraging T-Online's market leadership to grow advertising and e-commerce revenues
Systems integration	Creation of T-Systems as the second largest systems integrator in Europe
Local access	Expanding position as Europe's premier DSL provider

Source: Deutsche Telekom and author's interviews.

Table 4.13. Deutsche Telekom: corporate strategy 3: strategy by pillar

T-Mobile	T-Online	T-Systems	T-Com
Mobile telecoms	Internet services	Data/IP and systems solutions	Fixed network, telecoms infrastructure, CATV
Strategy	**Strategy**	**Strategy**	**Strategy**
Expand global mobile network and services	From access to access + content; expand European leadership	Provide integrated data and systems solutions	Increase network efficiency and speed, push DSL

Source: M. Fransman.

an analysis of Deutsche Telekom's recent acquisitions, VoiceStream is by far the company's largest target. It is no exaggeration to say that for the time being Deutsche Telekom's globalization strategy stands or falls by its success with VoiceStream.

Also in mobile, Deutsche Telekom is committed to pushing ahead with GPRS (general packet radio service), a so-called 2.5 generation mobile technology that allows for a higher rate of data transmission and, since information is sent in the form of packets, facilitates the 'always on' functionality that has helped to make DoCoMo's i-mode service so popular in Japan (as shown in Chapter 9).

In the field of the Internet, Deutsche Telekom will work hard to try and ensure increasing revenues from advertising and e-commerce. While the company is the major provider of ISDN (integrated services digital network) and ADSL (asymmetrical digital subscriber line) broadband access services in Germany, and although take-up of these services is growing rapidly, they remain 'commodity' services that can potentially, if the conditions are right, be provided by many network operators. Therefore, they are services for which there could be strong competition and low profit margins. This makes all the more attractive for Deutsche Telekom a strategy of moving up the value chain into higher profit margin areas such as advertising and e-commerce.

Finally, Table 4.12 also indicates Deutsche Telekom's determination to take advantage of the convergence of computing and telecommunications and reap profit by providing solutions based on systems integration. In the section below on T-Systems we will examine some of the problems that have confronted Deutsche Telekom in this area.

In Table 4.13, Deutsche Telekom's strategies are shown by 'pillar'.

GLOBALIZATION

As already noted, globalization constitutes a key plank in Deutsche Telekom's strategic armoury. By gaining access to markets outside Germany the company

Table 4.14. Deutsche Telekom: recent acquisitions

Company	Value (billion Euros)
VoiceStream	59.7
One2One	11.3
Powertel	6.9
Debis Systemhaus	4.4
MATAV (increase)	2.2
Media One	2.1
Club Internet	1.9
max.mobil (increase)	1.1
Slovak Telecom	1.0
Croatian Telecom	0.8
comdirect	0.7
SIRIS	0.7
Ya.com	0.6
Beta research	0.5
Total	93.9

Source: Deutsche Telekom, January 2001.

hopes it will be able to significantly increase its growth rate. This has been made clear by CEO Ron Sommer: 'One of the cornerstones of our strategy is internationalization, which will account for the lion's share of future growth in our four strategic business areas . . . we continued to focus our internationalization policy toward mergers and acquisitions and obtaining majority stakes in our subsidiaries. Apart from being able to consolidate our revenue in these endeavors, we are also assured of management control'.[5]

A snap-shot of the state-of-play regarding Deutsche Telekom's globalization strategy is provided in Table 4.14 that shows the company's recent acquisitions ranked according to value.

Several observations may be made on the basis of the information provided in Table 4.14. The first is the overwhelming importance Deutsche Telekom has attached to its acquisition of VoiceStream. As will be indicated shortly, this acquisition comes on the heels of several high-profile failures for Ron Sommer and Deutsche Telekom, failures that appeared to leave the company's globalization strategy rudderless. VoiceStream accounts for 64 per cent of the total amount shown in Table 4.14 committed by Deutsche Telekom to acquisitions.

Second, the next two acquisitions, in order of monetary value, are also in the field of mobile communications. These are One2One, the fourth mobile

---

[5] Deutsche Telekom, *1999 Milestones*, p. 4. Clearly, Deutsche Telekom's globalization strategy has been significantly slowed by the Telecoms Bust from March 2000 analysed in Chapter 1.

Table 4.15. Deutsche Telekom: new acquisitions and businesses by pillar

T-Mobile	T-Online	T-Systems	T-Com
Mobile telecoms	Internet services	Data/IP and system solutions	Fixed network
VoiceStream (US)	Club Internet (France)	Debis Systemhaus (Germany)	MetroHoldings (UK)
One2One (UK)	T-Online (Austria)		Siris (France)
Powertel (US)	Ya.com (Spain)		Matav (Hungary)
maxmobil (Austria)			Multilink (CH)
Ben (NL)			
Div. (E.Europe)			

operator in the UK, and Powertel, a mobile operator in the US. The high priority that Deutsche Telekom has attached to mobile communications, as evident in its allocation of investment resources to this area, is explained by the rapid growth in the company's revenue attributable to this source. As was shown in Table 4.5, while mobile in 2000 accounted for 22 per cent of Deutsche Telekom's total revenue, it was the fastest growing area in the company with mobile revenue growing at 95 per cent. Several of the other smaller acquisitions shown in Table 4.14 are also in the mobile area, mainly in smaller European countries.

In Table 4.15, further information is provided on Deutsche Telekom's acquisitions by business pillar.

Deutsche Telekom's Global Disasters

Having been appointed as Chairman of the Deutsche Telekom Board of Management in 1995, and having formulated the company's globalization strategy a few years later, Ron Sommer's globalization efforts ran into disaster in 1999. In 1999 he suffered a number of major setbacks. Having carefully nurtured a global alliance with his neighbour, France Telecom (Michel Bon, CEO of FT, was on Deutsche Telekom's Supervisory Board from 1998), Ron Sommer in 1999 made an opportunistic move to acquire the Italian incumbent, Telecom Italia. The problem was that Sommer neither consulted nor informed FT of his intentions before the move became public. This breach of confidence destroyed at a stroke Deutsche Telekom's alliance with FT (that also included Sprint, the third-largest US long-distance carrier).

In May 1999, however, it became clear that Deutsche Telekom had lost out to Olivetti, the Italian computer and communications company, in its bid for Telecom Italia. Sommer followed with an attempt to acquire Sprint. But this too failed in September 1999 when it was announced that WorldCom had agreed to buy Sprint, a move that in turn was foiled by US and European regulators on grounds of the competitive implications of the deal. Sommer then made another unsuccessful bid for Sprint. Sommer's eventually successful bid for VoiceStream, the US mobile operator already discussed in this chapter, therefore represented the next step in Deutsche Telekom's bid to become a credible player in North America.

For Deutsche Telekom these failures have been an undoubted setback and explain the importance the company attaches to its VoiceStream acquisition. For FT, however, there is compelling evidence that the break-down of its alliance with Deutsche Telekom has, if anything, assisted a more focused effort at Europeanization.[6] However, amongst the European incumbent telecoms carriers Deutsche Telekom stands alone[7] in attempting to establish itself as a genuinely global player. By contrast, BT and FT have made the decision to restrict their global ambitions to becoming European players while providing global network services to their large company customers. Their decision was influenced by a combination of the Telecoms Bust from March 2000 and their rising debt levels occasioned by acquisitions and the cost of third-generation mobile licences and infrastructure.

T-MOBILE

Deutsche Telekom's position in the German mobile market, through its T-Mobile pillar, is shown in Table 4.16.

As Table 4.16 makes clear, T-Mobile's market share in the German mobile market (39.4 per cent) is practically the same as the leader, Vodafone (formerly Mannesmann in Germany) with a share of 40.6 per cent. This compares with NTT DoCoMo's share of 58 per cent in Japan, France Telecom's share of 48 per cent in France, and BT's share of 25 per cent in Britain. The third and fourth players in the German mobile market, E1 and E2 (owned by BT), have far smaller shares at 14.7 per cent and 5.3 per cent respectively.

Deutsche Telekom's main international mobile interests are shown in Table 4.17.

Several important points emerge from Table 4.17. The first, and most significant, is Deutsche Telekom's relatively weak position outside Germany. In Europe the company's most important market outside Germany is the UK. However, in the UK T-Mobile's acquired subsidiary, One2One, is generally acknowledged as the weakest of the country's four mobile players. One2One

---

[6] This evidence is analysed in detail in the following chapter on France Telecom.

[7] The Spanish incumbent, Telefonica, is a partial exception to this statement with its considerable investments in Latin America where the linguistic affinity has been an important determinant of business opportunities.

Table 4.16. Deutsche Telekom: market shares in the German mobile market, mid-2000

Company	Market share (%)
D2 (Vodafone)	40.6
D1 (Deutsche Telekom)	39.4
E1 (E-Plus Mobilfunk)	14.7
E2 (BT/Viag Interkom)	5.3
Total market penetration:	58.5

Table 4.17. T-Mobile international group: main associated companies, 30 June 2000

Country	Company	DT percentage	Customer numbers (million)
Germany	T-Mobile	100	13.4
UK	One2One	100	6.0
Austria	Max.mobil.	100	1.8
Czech Republic	RadioMobil	60.8	1.0
Hungary	Westel	49	1.3

has a market share of 21 per cent compared to Orange's 24 per cent (owned by France Telecom), BT Cellnet's (now mmO2) 25 per cent, and leader Vodafone's 30 per cent. Significantly, Deutsche Telekom does not feature in the other major European markets, France, Italy, and Spain. In the US, as already noted, Deutsche Telekom is pinning its hopes on its acquisition of VoiceStream and Powertel announced in 2000.

Second, it is also clear from Table 4.17, that Deutsche Telekom has gone for majority control, and indeed for 100 per cent control, of its foreign mobile interests where possible. As mentioned earlier, this contrasts with Vodafone, NTT DoCoMo, and BT Cellnet that have all accepted that in some cases it may be necessary to begin with minority holdings.

Finally, Deutsche Telekom's emphasis on eastern Europe is also evident from Table 4.17, hardly surprising in view of the company's substantial activities in the former East Germany after reunification.

Organization of Deutsche Telekom's Foreign Mobile Assets

In order to more effectively organize its mobile assets in Germany and abroad Deutsche Telekom, on 1 January 2000, established T-Mobile International. The company's German mobile subsidiary, T-Mobile, was transferred to T-Mobile

International together with One2One in the UK, max.mobil in Austria, MTS in Russia, RadioMobil in the Czech Republic, and PTC in Poland. Each of these businesses in the countries outside Germany, as well as T-Mobile in Germany, report directly to the CEO of T-Mobile International. (See Figure 4.1 for an account of how T-Mobile International fits into Deutsche Telekom's general corporate organization.)

T-Motion

In March 2000, T-Mobile International and T-Online, Deutsche Telekom's Internet pillar analysed below, jointly created a mobile portal called T-Motion in order to facilitate mobile e-commerce services. T-Mobile International holds 60 per cent of the shares of T-Motion. Interestingly, T-Motion, intended to serve Deutsche Telekom's customers outside Germany, was created in London. At the beginning of 2001 T-Motion employed about 100 people. The company provides mobile Internet services using the WAP (wireless application protocol) standard.

In Table 4.18 the most popular services provided by T-Motion are shown.

Although this is a rapidly evolving area where consumer tastes have not yet stabilized, it is worth noting that information services are the most popular category. This contrasts strongly with NTT DoCoMo's i-mode, the most successful mobile internet service globally, where entertainment is the most sought after service.   .

T-ONLINE

T-Online, Deutsche Telekom's Internet pillar, is the largest Internet company in Europe. This is shown in Table 4.19 which gives subscriber figures for the top five Internet companies operating in Europe.

Table 4.18. T-Motion (Deutsche Telekom's mobile portal): most popular services

Service	percentage of all hits
Information (news, business news, weather)	25
Community (dating zone, chat, pride guide)	16
Search (phone directory, yellow pages, doctor search, jobs)	15

Source: Deutsche Telekom, January 2001.

Table 4.19. Europe's largest Internet companies, 2000

Company	Mother company, country	Subscribers (million)
T-Online	Deutsche Telekom, Germany	7.9
Tiscali	Independent, Italy	4.9
Wanadoo	France Telecom, France	4.5
AOL	Independent, USA	4.0
Terra Lycos	Telefonica, Spain	3.5

As Table 4.19 makes clear, three of the top five Internet companies in Europe have been established by incumbent carriers: T-Online by Deutsche Telekom (first), Wanadoo by France Telecom (third), and Terra Lycos by Telefonica (fifth).[8] Of the two independents, one is Italian (Tiscali which is second largest and which is based on the island of Sardinia), and the other is American (AOL, the largest Internet Service Provider in the world that has merged with Time Warner). This situation in Europe contrasts strongly with that in the US where it is new companies, such as AOL, Yahoo! and WorldCom (with its UUNet subsidiary), that have dominated the Internet.

Unfortunately, Deutsche Telekom does not publish a detailed breakdown of T-Online's revenues. The information that is available has to be pieced together but even then nothing like the full picture emerges. We do know, however, as shown in Table 4.5, that Deutsche Telekom's revenue from data communications (which presumably includes T-Online's traffic) increased by 21 per cent from 1999 to 2000, although revenue from this source contributed only 8.8 per cent to the company's total revenue. We also know that, in line with experience in other countries around the world, T-Online's subscribers spend more minutes each month accessing the Internet through T-Online. The figures are given in Table 4.20.

We also know that T-Online's revenues from e-commerce and advertising are growing rapidly. The data are provided in Table 4.21.

Although the data are not available to compare revenue per subscriber at T-Online with that of the other big five European Internet companies, there is information comparing Tiscali, the second largest Internet company in Europe, with AOL's worldwide figures. This is shown in Table 4.22.

We do know that T-Online's subscribers spend significantly less time on T-Online's site (including directly linked sites) than AOL's. In large part the reason is the additional amount of content available in AOL's 'walled garden', content that has been significantly increased by this company's acquisition of Time Warner. However, the figures in Table 4.22 for Tiscali and AOL have to be treated with caution because AOL's US customers tend to spend longer online than their European counterparts.

---

[8] Terra Networks, Telefonica's Internet company, acquired the American internet company, Lycos.

Table 4.20. T-Online: growth in minutes
per month per subscriber

Year	Period	Minutes/month/subscriber
1997	H1	190
	H2	205
1998	H1	225
	H2	245
1999	H1	300
	H2	350
2000	Q1–Q3	501

Table 4.21. T-Online: e-commerce
and advertising revenues, 2000

Quarter	Revenues (billion Euros)
Q1 2000	15.9
Q2 2000	17.4
Q3 2000	27.8

Table 4.22. Tiscali and AOL: advertising revenue
per customer, 2001

Company	Advertising revenue per customer ($)
Tiscali (in Europe)	1.40
AOL (worldwide)	4.70

Source: *Business Week*, 26 February 2001, p. 19.

T-Online, like T-Mobile, has recently begun to expand outside Germany although the company does not yet have a really significant market in other countries. Table 4.23 shows T-Online's main foreign subsidiaries.

T-SYSTEMS

T-Systems is the latest of Deutsche Telekom's business pillars, established in January 2001. T-Systems is the second-largest systems house in Europe for information technology and telecommunications, coming after IBM. The aim of T-Systems International GmbH is to take advantage of the demand by larger

Table 4.23. T-Online: major foreign subsidiaries

Subsidiary	Country	T-Online's share (%)
Club Internet	France	99.9
Ya.com	Spain	92.66
T-Online.at	Austria	100

On 17 April 2000 a 9.32 per cent share of T-Online was floated.

customers for integrated information and communications solutions based on systems integration. In Deutsche Telekom's words, 'The business strategy of the new Frankfurt-based service enterprise is centred on the convergence of IT/telecommunications technologies, products, services and networks and the e-business solutions evolving from this process'. Deutsche Telekom is extremely bullish about the prospects for T-Online. According to Josef Brauner, member of Deutsche Telekom's Board of Management in charge of sales and customer care, 'We expect [T-Systems] to contribute as much to Deutsche Telekom's total revenue... in five years time as from our current core business, the telephone service'. Deutsche Telekom hopes that by moving up the value chain into the customized solutions business it will be able to benefit from higher profit margins compared to those that are earned in 'commoditized' markets such as pure transport and Internet access.

In order to be able to become a leading contender in this market in Europe, however, Deutsche Telekom had to strengthen its competencies. Although it already had its own systems integration business, DeTeSystems, this was not enough. Accordingly, the company entered into negotiations with DaimlerChrysler to acquire its systems business, debis Systemhaus. Deutsche Telekom succeeded in taking a 50.1 per cent stake in debis Systemhaus that was already the second-largest such business in Europe after IBM even though most of its customers were in manufacturing rather than information and communications. In September 2000 Deutsche Telekom was given regulatory approval for the acquisition from the German authorities.

Having acquired debis Systemhaus, Deutsche Telekom set a tight schedule to integrate the company into T-Systems International. The result, however, was not what Deutsche Telekom expected, as is shown in detail towards the end of this chapter.

T-COM

As shown in Table 4.4, the contribution to Deutsche Telekom's revenue by T-Com, Deutsche Telekom's fourth business pillar, fell sharply from 78 to 67 per cent in the third quarters of 1999 and 2000 respectively. This reflected the decline in Deutsche Telekom's network communications business (shown in Table 4.5). T-Com, therefore, is the business area where Deutsche Telekom, like its incumbent counterparts in the other large industrialized countries, is feeling the squeeze of strong competition coupled with rapidly falling prices.

In part, T-Com's strategy is to compensate for this by increasing network speed and bandwidth and offering enhanced services such as Internet access via ADSL. The option of using Deutsche Telekom's substantial cable TV assets, however, was thwarted by the European Union's regulators in Brussels.

T-Com's ADSL Versus Cable

As a result of the Brussels ruling, Deutsche Telekom was forced to sell a majority stake in its nine regional cable TV companies. The sell-off began at the end of 2000 and into 2001 with 55–65 per cent of several regional companies being sold to Liberty Media Corp. (owned by AT&T) and the German cable company, Klesh & Co. With this deal and a separate agreement for Liberty Media to increase its holding in UnitedGlobalCom, the company that controls United Pan-Europe Communications NV, Liberty Media became Europe's biggest cable company.

However, Deutsche Telekom has retained at least a strategic 25 per cent plus one share holding in each of the nine regional cable companies which, under German law, give Deutsche Telekom strong veto rights, including veto over decisions to upgrade the cable networks of these companies. The details of the agreement between Deutsche Telekom and the purchasers, however, have not been made public. Neither is it clear how the German regulatory authorities are likely to view the obvious conflict of interest that exists here.

R&D

Deutsche Telekom's expenditure on R&D relative to the other Big Five incumbents is shown in Table 4.24.

Table 4.24. Deutsche Telekom: R&D expenditure

Telecoms operators	1999 R&D spend ($'000)	Sales ($m)	R&D percentage sales	R&D employee ($'000)
AT&T	550,000	62,391	0.9	2.3
BT	556,037	30,163	1.8	2.6
Deutsche Telekom	701,611	35,552	2.0	2.2
France Telecom	594,572	27,297	2.2	2.1
NTT	3,729,910	95,061	3.9	10.3

T-Nova—Deutsche Telekom's Innovation Subsidiary

As is clear from Table 4.24, Deutsche Telekom is right in the middle of the Big Five in terms of its R&D intensity—2.0 per cent of sales—and R&D per person. In July 1999 Deutsche Telekom implemented significant changes in its organization of R&D when it established T-Nova its 'innovation subsidiary'. The aim of T-Nova was to bring together under one roof its various units involved in R&D: the company's Technology Center, its five Development Centers, and the business activities of Berkom, its unit focusing on the development of future-oriented telecoms applications. In addition, T-Nova was put in charge of Deutsche Telekom's Multimedia Software GmbH Dresden, which develops software for multimedia, Internet, and intranet applications. Deutsche Telekom argues that 'The decisive point is that by bringing our research and development units together, T-Nova can speed up know-how transfer and act as a supra-regional communications and action management center for projects'. In 1999 4,400 people were employed in T-Nova, Multimedia Software GmbH Dresden, the central Innovation Management Unit, and the development division of T-Mobile.

The reorganization of R&D also saw T-Nova being given a more significant role to play in strategically significant development projects with a short-term payoff. For example, in 1999 T-Nova took over all the activities involved in both the development and the introduction of Deutsche Telekom's T-DSL services for broadband Internet access. T-Nova also took responsibility for the development of important software modules and technical systems integration for the company's SMILE system, a system that aims at optimizing customer care for residential and business customers.

The rapid increase in patent applications by Deutsche Telekom is shown in Table 4.25.

Table 4.25. Patent applications by Deutsche Telekom

Year	Number of patent applications
1993	39
1994	72
1995	158
1996	270
1997	353
1998	363
1999	454

Source: Deutsche Telekom, Annual Report, 1999, p. 24.

EMPLOYMENT

The change in employment in Deutsche Telekom over the period 1995, when the company was formed, to 1999 is shown in Table 4.26 by type of employment.

Several points are worth noting from Table 4.26. First, total employment has been declining steadily each year since the company was formed in 1995. This is the case despite steadily increasing revenue. Second, the number of employees classified as civil servants has declined sharply. However, in 1999 more than 40 per cent of total employees were still classified as civil servants. Third, the number of salaried employees has been increasing, although in 1999 there were still more civil servants than salaried employees. Finally, the number of wage earners—that is employees with less skills than the other two categories, declined slowly indicating that the company, like the Telecoms Industry as a whole, was becoming more skilled-labour-intensive.

DEUTSCHE TELEKOM: THE POLITICS OF CORPORATE REORGANIZATION

Deutsche Telekom, like the other incumbents in the Telecoms Industry and new entrants such as WorldCom and Qwest, has depended to a significant extent for the implementation of its strategies on acquisition. This is clear, for example, in the case of Deutsche Telekom's acquisition of VoiceStream, debis Systemhaus, and One2One. The debis Systemhaus acquisition was crucial in building up Deutsche Telekom's competencies in systems integration, as shown in the analysis of T-Systems. Similarly, the acquisition of One2One has been essential to bolster Deutsche Telekom's position in the European mobile market.

But acquisition is not always the panacea it is often made out to be and, as in the case of Ron Sommer's attempts to empower the CEOs of the company's four business pillars that we have already analysed, the intentions of company leaders are often frustrated by unforeseen events. As things turned out, the integration of debis Systemhaus, formerly a subsidiary of DaimlerChrysler,

Table 4.26. Deutsche Telekom: employment

Type	1999	1998	1997	1996	1995
Civil servants	71,123	83,848	91,542	105,419	114,430
Salaried employees	63,590	54,008	51,681	44,235	43,672
Wage earners	37,520	41,313	47,811	51,406	55,365
Total	172,233	179,169	191,034	201,060	213,467

Source: Deutsche Telekom, Annual Report 1999, p. 27.

Table 4.27. Deutsche Telekom: the politics of integrating acquisitions

On 1 January 2001 Deutsche Telekom launched T-Systems as Europe's second largest systems house for information technology and telecoms after IBM. T-Systems is one of Deutsche Telekom's four strategic 'pillars'. For some time Deutsche Telekom had its own systems integration business, DeTeSystems. However, the business was not particularly strong and lacked significant capabilities. In order to strengthen this business and position it more strongly in a market that was seen to have the potential to replace Deutsche Telekom's sagging core business, namely telephone services, the company decided to make a major acquisition. The target was debis Systemhaus, a former DaimlerChrysler subsidiary, which was the second largest systems house in Europe after IBM with almost half of its revenue coming from the manufacturing sector. In 2000 Deutsche Telekom took a 50.1 per cent stake in debis Systemhaus, getting regulatory approval from the German authorities in September 2000. A tight schedule for the integration of DeTeSystems and debis Systemhaus was planned, beginning on 1 October 2000 and intended to be completed by 1 January 2001. On 1 January 2001 T-Systems was duly launched. However, with the exception of one manager, the entire top management of debis Systemhaus resigned. As a consequence, more than one-third of the approximately 20,000 former employees of debis Systemhaus walked out during the integration process. A report in the German business press identified Ron Sommer's desire for 'absolute power' as the main reason for the problems experienced in integrating debis Systemhaus into the fabric of Deutsche Telekom. Deutsche Telekom insiders have their own interpretations, arguing that despite the difficulties T-Systems will become a major source of strength for the company.

presented far more problems than had been anticipated. A summary of this story is given in Table 4.27.

What are the implications of the debis Systemhaus story? The main implication for Deutsche Telkom is the cautionary conclusions that are implied from this attempt to grow quickly through using the company's shares as currency to purchase other companies. While most large and successful telecoms companies have used this mechanism to facilitate fast growth—and these include both incumbents as well as newer companies like WorldCom and Qwest—it is wise not to underestimate the costs of this approach. In those cases where the acquired company can continue its activities largely unimpeded, and simply be added as an additional 'module' to the acquirer's portfolio, success is more likely to result. However, in the instances where, like debis Systemhaus, closer integration of the human and physical resources acquired into the existing fabric of the company is necessary, success is likely to be harder to achieve. The lesson for Deutsche Telekom is that 'Going for Globalization' may very well be a more complicated and costly task than Sommer and his team often suggest. A significant increase in the cost side of the cost-benefit calculus may require some rethinking regarding the company's globalization strategy.

THE FUTURE OF DEUTSCHE TELEKOM

What does the future hold for Deutsche Telekom? It is possible, in answering this question, to construct two very different stories regarding the future of the company. We begin with the optimistic story.

The Optimistic Story

Deutsche Telekom's problems in the opening years of the new millennium will be rather short-lived and the company will soon recover. The company's huge indebtedness coupled with a significant fall in its share price, a problem common to most of the large incumbents and new entrants in the Telecoms Industry after the Telecoms Bust, will soon be overcome as a result of financial restructuring and a stabilizing of the bear stock market in telecoms and technology shares.

Into the medium term Deutsche Telekom will be able to make much of its strong brand name in Germany and the full range of integrated information and communications services that it is able to deliver. In Germany the company will be able to take advantage of its unparalleled networks that provide greater access to more customers than any other competitor. Not only large companies, but also other types of customers, will turn to it as the single provider of choice for the most important and increasingly essential telecoms-related services. With increasing capabilities, not only in the commodity business of carrying bits, but also in the higher value-added areas of systems integration and info-communications solutions, Deutsche Telekom will be able to become more profitable and grow more rapidly. Being an integrated network operator and service provider, Deutsche Telekom will be able to realize important synergies between its various networks and services. For example, T-Mobile and T-Online will be able to offer integrated mobile Internet services that take full advantage of both the mobile and the fixed sides of the telecoms business. Together with T-Systems, business customers will be able to draw on the full range of competencies that will provide them with profit-enhancing solutions to their business problems. Similarly, residential customers will be provided with new services and improved old services in their increasingly networked homes.

The company's globalization strategy will also start to succeed. Particularly in the mobile area, Deutsche Telekom's acquisitions both in Europe and the US will give the company a strong international positioning, making it a major competitor to the main mobile players such as Vodafone, NTT DoCoMo, and FT/Orange. Mobile will continue to be one of the fastest sources of growth in revenue and profitability for the company and this will be enhanced by significant demand for the new range of services that will become available with 2.5- and 3-generation networks.

In the growth area of broadband Internet access and broadband services Deutsche Telekom will also be able to make the most of synergies between its four business pillars. T-Com will provide the ADSL connections to offices and homes that will also generate new business opportunities for T-Online, T-Systems, and T-Mobile. Using the strong German market as both a learning ground and a springboard, Deutsche Telekom's four pillar businesses will increasingly be able to extend their services to the other major European countries and to eastern Europe where, geographically and historically, the company is particularly well-placed. In this effort, T-Online will be able to leverage its position as by far the largest Internet Service Provider in Europe.

In summary, the advantages referred to will provide Deutsche Telekom with a strongly competitive position that will allow the company to survive and thrive in competition with the strongest global competitors, particularly in Europe but also, in areas such as mobile, in North America.

The Pessimistic Story

There is, however, a more pessimistic story that may be constructed from the evidence that has been provided in this chapter on Deutsche Telekom. In contrast to the optimistic story, the pessimistic one would emphasize the following factors.

The main problem, according to the pessimistic story, is that in the medium to longer term there are major threats to Deutsche Telekom's revenue and profits stream. The company's problem of huge indebtedness in early 2001—seen by most analysts as being caused by a combination of acquisition costs, third-generation auctioned mobile licences, and infrastructure costs—only serves to obscure more fundamental problems.

These fundamental problems can be simply captured, as they are in Table 4.28 (which should be read in conjunction with Table 4.5).

In order to highlight Deutsche Telekom's problems we need only to look at two of the segments shown in Table 4.28: network communications and mobile communications. In 1999 network communications accounted for 47 per cent of the company's entire net revenue, leading Deutsche Telekom to acknowledge in its 1999 Annual Report that this segment 'remains the company's largest revenue driver...[and] is still the decisive segment contributing to the Group's results' (p. 131). However, the contribution of network communications to growth in revenue and profitability is declining rapidly. In 1998 this segment contributed 58 per cent to net revenue, the contribution falling, as we have just seen, to 47 per cent only one year later. Between 1998 and 1999, as shown in Table 4.28, income before taxes from network communications fell by

Table 4.28. Deutsche Telekom: net revenue and income before taxes by segment, 2000 (bold underlined, percentage of total net revenue in brackets), 1999 (bold), and 1998

Segments	Net revenue (Euros million)	Income before taxes (Euros million)
Network communications		**15,051 (36%)**
	16,737	**2,276**
	20,531	4,796
Carrier services		**3,983 (10%)**
	2,884	**440**
	1,611	589
Data communications		**3,352 (8%)**
	2,828	**104**
	2,536	(61)
Mobile communications		**9,253 (23%)**
	3,919	**1,033**
	3,061	560
Broadcasting and broadband cable		**1,861 (5%)**
	1,917	**(86)**
	1,804	(329)
Terminal equipment		**1,036 (3%)**
	1,207	**19**
	1,382	(114)
Value-added services		**1,802 (4%)**
	1,903	**(152)**
	2,051	(182)
International		**2,175 (5%)**
	2,863	**(339)**
	1,322	200
Other segments	**1,122**	**(408)**
	772	(384)
Group		**40,939 (100%)**
	35,470	**2,944**
	35,144	5,100

Source: Deutsche Telekom, Annual Reports 1999, 2000.

53 per cent, from 4,796 million Euros in 1998 to 2,276 in 1999. But in 1999 network communications generated no less than 77 per cent of the Deutsche Telekom group's total income before taxes. Furthermore, as Table 4.5 shows, the revenue generated by network communications in the first three quarters of 2000 was down by 10.7 per cent compared to a year earlier.

This raises two key questions. First, why has revenue and profitability from network communications fallen so much? Second, what will replace this most important source of revenue and profitability?

The answer to the first question is straightforward. As Deutsche Telekom succinctly puts it in its 1999 Annual Report, 'the main causes of the reduction in revenue and income [from network communications] are the price cuts... and the impact of competition' (p. 131). In short, it is precisely here, at the heart of Deutsche Telekom's revenue and profit, that competition is biting—and hurting.[9]

Where will Deutsche Telekom find relief from this haemorrhaging of its revenue and profits (i.e. the second question)? The answer so far is from the rapidly expanding mobile communications segment. In its 1999 Annual Report, Deutsche Telekom acknowledges that 'along with network communications, mobile communications... is a major contributor to the Group's results' (p. 132). In 1999, as shown in Table 4.28, mobile communications accounted for 35 per cent of the Group's income before taxes. Moreover, it is apparent from Table 4.5 that over the year ending at the end of the third quarter in 2000, revenue from mobile communications increased by 95 per cent, by far the fastest growing segment in the whole of Deutsche Telekom.

But how long will mobile continue to bail the company out? This is the key question. Deutsche Telekom in Ron Sommer's public addresses makes much of the company's large and growing number of mobile subscribers, not only in Germany (where the company runs a close second to Vodafone) but also in expanding markets abroad. The latter include the UK and, with the success of the VoiceStream acquisition, the US. However, it is by no means clear that Deutsche Telekom will be able to continue growing revenue and profits at the same rate from mobile. Saturation of existing mobile markets—such as 'plain old mobile voice'—is a looming reality while the ability of new services, including the much-vaunted third-generation services, to fill the gap is subject to a significant amount of uncertainty. In short, it is by no means clear how much consumers will be willing to pay for the new services.

As we saw earlier, Deutsche Telekom argues that its newest business pillar, T-Systems, will make a major contribution to the growing hole in revenue and profit caused by the ailing network communications segment. It may well be, however, that this forecast is overly optimistic. T-Systems may be the second-largest information and communications system house in Europe after IBM. However, the implications may not be rosy for Deutsche Telekom. To begin with, size alone does not necessarily generate a competitive advantage. Secondly, as was mentioned earlier, debis Systemhaus, which forms the heart of T-Systems, was a DaimlerChrysler subsidiary and most of its customers were in the manufacturing sector. It is not yet clear how well T-Systems' capabilities, including marketing, will be applied in other sectors. Third, as we have seen, Deutsche Telekom has encountered serious problems in integrating debis Systemhaus and it is not certain how long it will take to resolve these problems. Fourth, the area of information and communications systems integration is

[9] The fundamental causes behind these trends were examined in Chapters 1 and 2.

hotly contested by many strong competitors that include computer hardware and software companies as well as telecoms equipment and software suppliers. T-Systems' performance against competitors of this calibre cannot be taken for granted. For all these reasons it must be concluded, taking a pessimistic view, that T-Systems may not be able to fill the revenue and profit gap left by Deutsche Telekom's declining network communications segment.

In the pessimistic view, these questions about revenue and profitability must be added to other structural weaknesses that limit Deutsche Telekom. One of the most important of these is the company's weakness in many of the world's major telecoms markets. In Europe, Deutsche Telekom is very weak in France, Italy, and Spain. In North America the company's hopes stand or fall by the VoiceStream and Powertel acquisitions. In Japan and Asia it has little credible positioning.

A possible further problem is that what Deutsche Telekom sees as one of its most important competitive advantages may turn out to be far less significant than the company hopes. Specifically, Deutsche Telekom makes much of its integrated corporate structure and its ability to provide a full range of telecoms services, from pure transport to integrated information and communications solutions. However, while there may be potential economies of scope to be earned, the opposite side of the same coin is the very complex corporate structure that Deutsche Telekom has created in order to deliver this wide range of services. While the decomposition of the company into four business 'pillars' looks, on paper, as if it might simplify the complexity of the company's organization, the cautionary tales that have been recounted in this chapter regarding the problems Sommer and his colleagues have encountered with the top management of T-Online and with the integration of debis Systemhaus suggest that problems of corporate size and complexity may continue to plague Deutsche Telekom.

In turn, this may open the door to 'F4 Competitors'—companies that, being much smaller, are more focused, flexible, fast, and flat than Deutsche Telekom, and therefore able to eat into its lucrative markets. Such nimble competitors may be able, through agreements with complementary companies, to take advantage of many of the 'synergies' that Deutsche Telekom is attempting to reap at the cost of a large and complex corporate organization.

To conclude, therefore, it is certainly possible, on the basis of the existing evidence, to construct a pessimistic scenario that will shape Deutsche Telekom's future.

CONCLUSION

The aim of this chapter has been to analyse the prospects for Europe's largest incumbent telecoms company, Deutsche Telekom. Based on the detailed evidence that has been closely examined it was concluded that it is possible for two very different stories to be constructed regarding the company's longer-term future. The first story, the optimistic scenario, emphasized the potential for growth in revenue and profit that exists as the information and communications revolution continues to unfold into the twenty-first century and the ability of Deutsche Telekom to take advantage of this potential. However, the second story—the pessimistic scenario—while not denying the extent or importance of the continuing information and communications revolution, questioned the ability of Deutsche Telekom to profit significantly from this revolution. The absence of significantly above-average longer-term profitability of companies in the electricity industry, despite the undoubted significance of the 'electricity revolution' in the early twentieth century, provides sobering food for thought for the pessimistic story. The question is whether at least some telecoms companies will be able to generate the scarcity rents that are always the source of above-average profitability within the context of the evolving global Telecoms Industry. And, if some are, whether Deutsche Telekom will be amongst them. One thing we can say with certainty is that time will tell.

5

France Telecom: France's Europeanizing Incumbent

In Chapters 1 and 2 the main forces driving the Telecoms Industry in the Internet Age were examined. How has France Telecom (FT), France's incumbent telecoms operator, coped with these forces in its attempt to survive and prosper? This is the overall question tackled in this chapter.

INTRODUCTION

France Telecom presents a paradox. In 2001 France Telecom was 54 per cent owned by the French government. Furthermore, it was only on 1 January 1998 that, under a directive of the European Union, FT faced full competition, some thirteen years later than AT&T, BT, and NTT. Under these circumstances conventional wisdom might predict that France Telecom, under the influence of the old PTT monopoly culture and lacking the pressure to change induced by competition, would perform badly. However, this has not been the case. As we will shortly show, FT has been one of the best performing of the Big Five incumbent network operators.

How is this paradox to be explained? This is the main question that will be tackled in this chapter through a detailed analysis of the evolution of this company.

PERFORMANCE OF FRANCE TELECOM: AN OVERVIEW

How should the performance of a telecoms company be measured? One candidate for performance indicator is the performance of the company's shares.

The author would like to acknowledge information received during interviews with several senior executives from France Telecom (FT). They, however, are not responsible for either the information or analysis provided here. This chapter has also benefited from Jackie Krafft's work on FT.

This indicator is based on the assumption that a company's share price reflects assessments of the present value of the future earnings that the company is expected to earn. These assessments about future earnings are made by the buyers and sellers of the shares who make their buy and sell decisions on the basis of information assembled, digested, and analysed by financial institutions. Taking all this information into account (information that may be contradictory and incomplete) buyers and sellers buy and sell the company's shares, thus driving the share price.[1]

One share price-based measure of performance is the absolute change in the company's share price. An increase in the share price over time reflects the judgement that the company's future earnings are expected to rise with the result that the present value of the future earnings, all other things equal, will rise. Another share price-based measure of performance is the change in the price of the company's shares relative to the share price of similar companies. This relative measure of performance measures not only how well the company has done compared to its own past performance, but also how well it has done compared to other similar companies in similar conditions.[2] Both these measures are shown in Table 5.1.

Table 5.1 shows that the decrease in FT's share price for the year to February 2001 was practically the same as that of four of the Big Five incumbents, Vodafone, and two of the three equipment suppliers. However, over the five-year period FT performed better than the other incumbents. For the rest of 2001 FT's relative share performance declined somewhat due to its higher level of indebtedness.[3] Nevertheless, FT remained one of the better performers amongst the incumbents. Over the five-year period, FT was significantly outperformed by Vodafone (which benefited from being entirely based in the high-growth mobile industry) and by two of the three equipment suppliers.

There are, however, shortcomings in the use of relative share price changes as an indicator of performance. The most important of these is that there are national influences on share price movements with the result that changes in share prices are not only a reflection of a company's expected future earnings. For example, Vodafone's share price is influenced by the fact that the company is the largest in the *Financial Times* FTSE 100 index with the result that investment funds attempting to track this index must hold Vodafone shares.

---

[1] Later in this chapter other performance indicators, such as market share, are used.

[2] Of course, external conditions are not always similar for different companies, particularly when these companies are from different countries. For example, macroeconomic conditions may differ from country to country and to some extent share price differences may reflect these differing conditions. This is a problem, for instance, in comparing NTT's share price with those of other telecoms companies during the 1990s when the Japanese economy was the only one to experience a prolonged downturn. These kinds of shortcomings must be kept in mind in using share price data to compare performance.

[3] In May 2001 FT's total indebtedness amounted to Euros 62.0 billion while that of Deutsche Telekom was Euros 56.0 billion and BT Euros 45.0 billion. The total revenue of the three companies was $31.3 billion for FT, $38.1 billion for Deutsche Telekom, and $32.2 billion for BT.

Table 5.1. France Telecom: relative share price performance[4]

Company	1-year change in share price (to 24/02/01) (%)	5-year change in share price (to 24/02/01) (%)	EPS (Mar. 2000)	P/E
Big five incumbents				
AT&T	−59.39	−29.64	0.53	12.60
BT	−54.99	+58.19	1.14	29.30
Deutsche Telekom	−68.90	+6.05	0.01	N/A
FT	−54.67	+76.17	N/A	29.90
NTT	−56.62	−14.74	N/A	49.2
Mobile leader				
Vodafone	−53.47	+278.62	N/A	55.5
Equipment suppliers				
Cisco	−55.12	+454.96	0.14	39.70
Lucent	−76.22	+73.11	0.25	−22.30
Nortel	−57.29	+254.49	0.12	26.20
Nasdaq	−44.16	+111.56		

Notes: EPS = earnings per share; P/E = price earnings ratio.

Source: http://quote.bloomberg.com 24/02/01.

Changes in national interest rates and macroeconomic conditions may also influence a company's share price relative to that of competitors in other countries. Accordingly, in assessing FT's performance it is necessary to include other measures of performance in addition to share price performance.

Perhaps one of the best additional indicators of performance is market share in the telecoms company's home market. In the case of FT (and Deutsche Telekom) the adequacy of this measure must be qualified to take account of the fact that full competition began officially only on 1 January 1998 under the directive of the European Union, although in many key markets, such as mobile, competition began years earlier. Despite this qualification, market share in the domestic market remains a useful indicator of performance.

How well has FT performed according to this indicator? Part of the answer is given in Table 5.2.

As Table 5.2 shows, FT is the market leader in France in the three main growth areas in the Telecoms Industry, namely mobile, Internet, and data. Not only this, but the company's market share is high, 49 per cent in mobile and 40 per cent in Internet (figures are not available for data). This compares very favourably with AT&T and BT. In AT&T's case the company's performance in mobile has been weak relative to national competitors such as Verizon Wireless and Sprint PCS. In the area of the Internet companies such as AOL and Yahoo! have significantly outperformed AT&T. In mobile, BT's performance has been exceeded by Vodafone and Orange while in the Internet area the company so

---

[4] The same diagram also appears as Table 4.1 in the last chapter on Deutsche Telekom.

Table 5.2. France Telecom: market share in France in the major growth areas

Business area	Market position	Market share
Mobile	Leader	49%
Internet	Leader	40%
Data	Leader	n/a

Source: France Telecom Annual Report, 1999, p. 27 and p. 22.

far has made little impact. Although FT's 49 per cent market share in mobile pales into insignificance compared to NTT DoCoMo's 60 per cent share, and although FT's share is slowly declining due to domestic competition, the company's dominance of the French market is unchallenged.

It may be concluded, therefore, that according to the domestic market share measure of performance FT's performance has been very good. The rest of this chapter seeks to explain this performance.

HISTORICAL BACKGROUND

As with the other European telecoms incumbents, telecoms services after the war were provided by a specialist ministry that also dealt with post. In the 1970s in France the *Direction Generale des Telecommunications* (DGT) was established in the Ministry with specific responsibility for telecoms infrastructure and services. In 1988 the name of DGT was changed to FT. On 1 January 1991 FT was legally separated from the Ministry. However, the French government continued to impose public-interest obligations on FT (see later for the obligations in the area of R&D).

In December 1996 laws were passed introducing competition into the French telecoms market and providing for the sale of FT shares to the public while retaining a majority shareholding by the French government. Furthermore, the ministry of telecommunications has the right to block or impose conditions on any proposed divestiture by FT of telecoms infrastructure necessary for the company to maintain its public-sector obligations. At the same time the legislation made provision for an independent regulatory authority called the *Autorite de Regulation des Telecommunications* (ART). Until this time it was the ministry of telecommunications that directly regulated FT through its *Direction Generale des Postes et des Telecommunications* (DGPT). On 1 January 1998, under a European Union directive, all the French telecoms markets, like those of the other European countries,[5] were opened fully to competition.

---

[5] Several countries were, however, given a little more time to comply with this directive.

Since 20 October 1997 FT's shares have been traded on the Paris Bourse. The company soon became the most valuable company quoted on this stock exchange. In 1998 the French government owned 75 per cent of the company while in 1999 the proportion went down to 63.2 per cent. At the beginning of 2001 the French government owned 54 per cent of FT.

CORPORATE ORGANIZATION 1996–2000

Since 1991 when FT was separated from the Ministry of Telecommunications many reorganizations were undertaken, particularly since 1996 by which time it had become clear that the company would face strong competition in all its markets. FT's organizational structure from 1996 to 2000 is shown in Fig. 5.1.

As shown in Fig. 5.1, FT was organized into five divisions. The first of these was the Network Division that was in charge of the company's domestic and international telecoms networks as well as the procurement of equipment and services for these networks. The next two divisions were customer-facing divisions that used the networks in order to provide services to customers, the Residential and Small Business Division and the Large Business Division respectively. The former division was responsible for mobile and multimedia services as well as for payphones and calling cards. The latter division dealt with data transmission, network services, broadcasting, and Global One, FT's

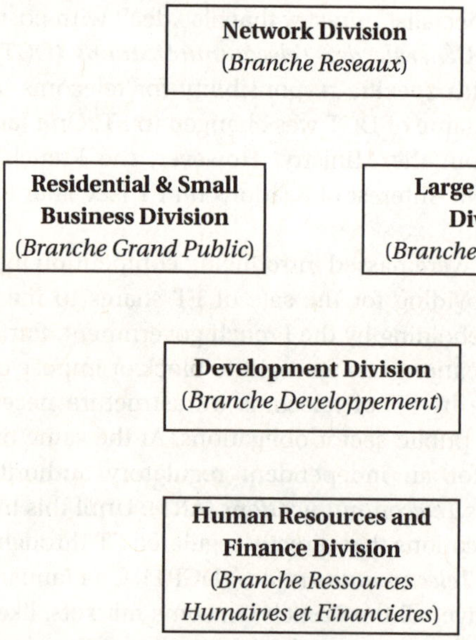

Fig. 5.1. France Telecom: organizational structure, 1996–2000.

Source: M. Fransman.

strategic partnership with Deutsche Telekom and Sprint. Finally, there were two support divisions. The Development Division dealt with FT's overall strategy, ran its R&D activities and information systems, and was responsible for domestic and international investment. The Human Resources and Finance Division's responsibilities are clear from the division's title. Within each division there was a decentralization of decision-making and financial accountability to operating units. At the beginning of 2001 there were some 400 operating units in FT.

Autonomous Subsidiaries

In addition to the five divisions described in Fig. 5.1 FT has established a number of autonomous subsidiaries. These include, Wanadoo, the company's Internet subsidiary which was originally established in 1996, and New Orange which was set up in 2001 after FT acquired the British mobile operator, Orange, and merged it with its own mobile operator, Itineris, and its other wireless interests. In July 2000 10 per cent of Wanadoo was floated and in February 2001 15 per cent of New Orange was floated. Wanadoo and New Orange are analysed in more detail in the sections below dealing with the Internet and mobile communications in FT.

CORPORATE ORGANIZATION FROM 2001

An overview is presented in Fig. 5.2 of FT's corporate organization from 2001.

As shown in Fig. 5.2, FT is divided into nine 'units' under the control of the Executive Board which consists of the Chairman and CEO, Michel Bon, the head of Corporate Communications, and the heads of the nine units.

The nine units, in turn, may be subdivided into three groupings. The first grouping consists of four units that are defined according to service/technology/network or customer segment. These are Mass Market Fixed Services, Mass Market Internet, Mobile, and Large Customers. The second grouping consists of two units that provide 'infrastructure' (or complementary assets) for the units in the first grouping. The second grouping consists of Networks and Distribution (France). Finally, the third grouping contains three units that 'support' the activities of the other six units. These are Development, Finance, and Human Resources.

Table 5.3 provides a more detailed description of the activities of each of the units.

Each of the nine units is organized as a separate profit and loss centre. Units pay one another in return for inputs received. For example, the Large Customers unit distributes/sells mobile services to large customers. In return, the Large

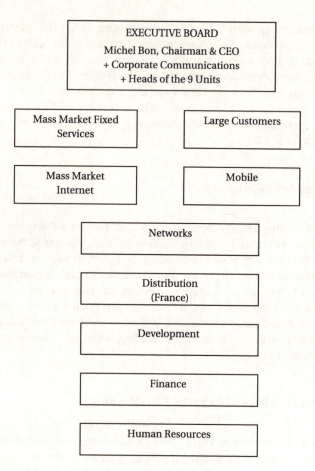

Fig. 5.2. France Telecom: organizational structure from 2001.

Source: M. Fransman.

Customers unit receives a commission from the Mobile unit. As far as possible market prices are used to determine cross-unit payments. In this way market prices are used as a coordinating mechanism within FT's internal organization.

Since the details in Table 5.3 on FT's nine units are self-explanatory, nothing more will be added here. In a later section information is provided on how FT organizes its international activities and how this organization relates to the corporate organization analysed in Fig. 5.2 and Table 5.3.

MAIN BUSINESSES

For reporting and accounting purposes, FT divides its activities into five business areas. FT's five business areas are shown in Table 5.4.

Table 5.3. France Telecom's nine units, 2001

Unit	Description
Mass Market Fixed Services	Mainly fixed voice services
Large customers	Large customers in France and abroad. Includes Global One and Equant. Includes voice services in France. Includes 'wholesale', selling network services to corporate customers and operators
Mass market Internet	Includes Wanadoo and its interests overseas (such as Freeserve in the UK)
Mobile	Includes Itineris in France, Orange in the UK, and FT's other mobile interests, now included in New Orange
Networks	Includes all FT's networks
Distribution (France)	Includes all FT's distribution assets (e.g. mobile outlets) in France
Development	Includes R&D, innovation, and strategy (the latter incorporating: strategic planning, strategic studies, international strategy, business intelligence, management, and competencies). Strategy also includes a strategy unit to identify international growth opportunities and oversee the consistence of development plans for FT's four regional zones
Finance	Finance, including the international finance unit created to coordinate financing of all international operations
Human Resources	Human resources

Source: M. Fransman.

Table 5.4. France Telecom's five business areas

Fixed telephony
Wireless
Internet and multimedia
Broadcasting and cable TV
Data

Source: France Telecom Annual Report, 1999.

Later in this chapter some of FT's activities in the fastest-growing business areas will be examined in more detail, namely mobile and the Internet.

Distribution of Revenue by Business Area

In Table 5.5 a slightly more detailed categorization of FT's business areas is used in order to analyse the distribution of the company's revenue.

Table 5.5. France Telecom: distribution of revenue by business area

Business area	1998 (% of total revenue)	1999 (% of total revenue)	mid-2000 (% of total revenue)
Fixed telephony	56.4	50.3	43.6
Mobile telephony	11.3	14.6	16.4
International	9.3	12.8	20.3
Leased lines and data	8.1	7.8	7.5
Information services	5.0	5.3	5.2
Broadcasting and CATV	4.0	3.6	3.2
Equipment	4.8	4.4	2.9
Other	1.1	1.2	1.0

Source: France Telecom, Annual Reports.

Several important conclusions emerge from Table 5.5. First, as with the other diversified telecoms network operators, FT's fixed telephony revenues, although providing the biggest single source of revenue, are declining rapidly in proportional terms. While in 1998 they provided 56.4 per cent of the company's total revenue, by mid-2000 this figure had declined to 43.6 per cent. The second important conclusion is that international operations are the second most significant contributor to total revenue. In 1998 these activities contributed only 9.3 per cent of total revenue, but by mid-2000 this had increased to 20.3 per cent. As is clear from Table 5.5, this source of revenue is growing more rapidly than any other. This is a reflection largely of FT's acquisitions and investments outside France, such as its acquisition of Orange and NTL (cable) in the UK, Mobilcom in Germany, and Global One (its former joint venture with Deutsche Telekom and Sprint). These acquisitions will be considered in more detail later.

The third conclusion to emerge from Table 5.5 is that mobile telephony (i.e. within France) is the third most important contributor to FT's total revenue. After international it is the second most rapidly-growing revenue source. This is in line with the global phenomenon of rapidly diffusing mobile telephony services.

Fourth, it is important to note that none of the other business areas is making a significant contribution to the growth of FT's revenues. One might have expected that leased lines and data would have registered a greater proportional contribution to total revenue. As the incumbent with a continuing monopoly over many parts of the French telecoms network, particularly local access, FT leases lines somewhere or other to most of the new entrants. Furthermore, data has become increasingly important as a proportion of total telecoms traffic partly as a result of the growth of the Internet. Yet while the contribution of leased lines and data to total revenue was 8.1 per cent in 1998, this figure had declined to 7.5 per cent by mid-2000. Although in absolute terms revenue from this business area increased over the period, competition in data and the regulation of leased lines charges limited the proportional contribution of this area to FT's total revenue.

While the contribution of information services was 5.0 per cent in 1998 it increased only slightly to 5.2 per cent in mid-2000. The contribution of broadcasting and CATV actually declined, from 4.0 per cent to 3.2 per cent. Significantly, equipment decreased from 4.8 per cent to 2.9 per cent, reflecting the continuing process of vertical specialization with specialist equipment suppliers such as Alcatel, Ericsson, and Nortel providing more and more of the industry's equipment needs.

Performance by Business

Further light is thrown on the performance of FT by business area in Table 5.6.

In Table 5.6, fixed telephony is lumped together with all FT's other services except for mobile in France and international operations. The table shows that although fixed telephony and other services in France contributed 79 per cent to the company total sales in 1999, there was negative growth in revenue from this segment between 1999 and 2000 of −0.3 per cent. On the other hand, while international operations contributed only 13 per cent of total sales in 1999, revenue from this source was growing at a rapid rate of 124 per cent. While mobile services in France contributed 15 per cent of the company's total sales in 1999, its growth rate over the two years was 47.4 per cent.

What conclusions regarding FT's future prospects can be inferred from Tables 5.5 and 5.6? To begin with, it is clear that FT is looking largely to international markets and to mobile markets in France to replace falling proportional revenues from fixed telephony. The company's future growth and prospects will accordingly be determined largely by changing conditions in these markets.

It is clear that the phenomenal growth rate from international operations, as already noted, is a reflection of FT's acquisition and investment spree in 2000,

Table 5.6. France Telecom: performance by business segment, 1999–2000*

Business segment	Percentage of total sales (1999)[†]	Percentage change (1999–2000)*
Fixed-line telephony and other services in France	79	−0.3
Mobile telecoms in France	15	47.4
International	13	124.0

Notes: [†]Does not total to 100 per cent due to eliminations.
*Based on first half year revenue figures.

Sources: France Telecom Annual Report, 1999, p. 92; France Telecom Press Release, 6 September 2000.

boosted by a strong share price until the first quarter of that year. FT's acquisitions and investment will continue, helped by the fact that despite the recession in telecoms shares in general from the second quarter of 2000, the company has managed to outperform the telecoms sector as a whole. However, it is obvious that revenue from international operations cannot continue to grow at this extremely rapid rate. Once FT has obtained its foothold in the other major European countries—particularly the UK, Germany, and Italy—opportunities for further viable investments will decrease. Moreover, competition in these countries will increase, both from their incumbents and the host of new entrants driven not only by local entrepreneurs but also from abroad, particularly the US. For these reasons the profitability from international operations cannot be assured.

The same applies to mobile. While the surge in demand for second-generation (Global System for Mobile Communications (GSM)) mobile telephony has provided both FT and its competitors with a significant source of growth and profitability, as it has mobile operators all over the world, there is an important degree of uncertainty surrounding the growth in demand for third-generation mobile services. Accordingly, FT's ability to earn a reasonable rate of return from its investment in third-generation mobile networks and auctioned licences remains to be established. This issue is taken up again in the section below dealing with FT's mobile communications.

EMPLOYEES

A further indication of FT's successful performance is provided by the growth of its employment. The data are shown in Table 5.7.

As can be seen from Table 5.7, FT's total employment has increased each year from 1997 to 2000. This period includes the time when the company's shares were first floated on the Paris Bourse and competition began to bite. While the increase in employment is also a reflection of the company's acquisitions and

Table 5.7. France Telecom: total number of employees

Year	Number
1997	165,042
1998	169,099
1999	174,262
2000*	188,830

Note: *First half 2000.

Source: France Telecom, Annual Reports.

investment, both in France and abroad, the numbers set FT against most of the other Big Five incumbents whose employment has fallen significantly over the 1990s. FT's employment figures, therefore, provide a further indication of the company's relatively good performance.

INTERNATIONAL BUSINESS STRATEGY

In the previous section FT's organizational structure and performance were examined. In analysing the company's performance it was seen that mobile telephony in France and international activities are key determinants of future growth and profitability. We now turn to examine FT's international business strategy as it existed at the beginning of 2001. This strategy is summarized in Table 5.8.

Several important conclusions emerge from Table 5.8. The first is that FT has abandoned any attempt (if such an attempt ever existed) to become a leading global player, preferring to specialize in the European market. This is important and distinguishes FT from AT&T, Deutsche Telekom, and NTT, all of whom, publicly at least, claim to have serious longer-term global ambitions. However, BT, like FT, is being forced by a combination of high debt and a falling share price into taking a similar Euro-centric position.

However, the second conclusion highlights the importance of distinguishing between genuinely global players who are contesting the major global markets and players who, while not present in strength in all the major markets, are nevertheless still offering global services to large business customers, particularly multinational companies. As France Telecom's second objective in its international business strategy, shown in Table 5.8, makes clear the company is to be included amongst the latter players. This is evident through FT's role in Global One, its former alliance with partners Deutsche Telekom and Sprint (considered later). Through Global One FT intends to offer mainly European-based large companies (some of whom may be headquartered elsewhere) access to competitive global services. While this will require an international network, it will not necessitate establishing a strong competitive position in the

Table 5.8. France Telecom's international business strategy

Objective
Become the leading pan-European operator.
Propose global services for business customers [in Europe].
Establish positions in emerging markets with strong growth potential.

Source: France Telecom News, 10 January 2001.

major markets outside Europe. In this sense it may be said that while there is an important global dimension to FT's pan-European strategy, the company is not aiming to be a fully globalized player.

The third conclusion to come from FT's international business strategy is the emphasis that the company is giving to 'emerging markets with strong growth potential'. For FT these markets are to be found largely in eastern Europe and, to a lesser extent, in those parts of the world where France has a historical interest such as French-speaking West Africa. This point will be highlighted later when a more detailed account is given of the new organizational structure put in place by FT to further its international ambitions.

France Telecom's Shifting Relationship with Deutsche Telekom

In order to fully understand FT's international business strategy it is necessary to comprehend its problematical relationship with Deutsche Telekom.

In the 'Age of Alliances'—when all the world's major incumbents, with the notable exception of NTT, stitched together global partnerships—FT teamed up with Deutsche Telekom and Sprint of the US, forming Global One in January 1996. The size of the companies, their dominant positions in the first and second largest markets in Europe, the contiguity of their operations in their home countries which created possibilities for synergies, and the political affinity of France and Germany in the European Union, seemed to provide a reasonable logic underpinning the alliance. Furthermore, the alliance would create a central European bulwark against the competing alliances established by the likes of AT&T and BT. Sprint, as the third largest US long distance company after AT&T and MCI, would provide a toehold in the crucial US market. The benefits of cooperation between the three companies were cemented by a 2 per cent cross shareholding as well as the participation of Deutsche Telekom's CEO, Ron Sommers, on FT's Board of Directors and Michel Bon, FT's CEO, on Deutsche Telekom's Supervisory Board.

Alas, this logic would soon prove to be less convincing than its implementers supposed. Not surprisingly, the rocks on which the alliance foundered were the rocks of self-interest, the same rocks that had sounded the death-knell for countless numbers of alliances over the ages. More specifically, it was the promised land of rich pickings close to home that tempted FT's partner, Deutsche Telekom, to break ranks and opportunistically pursue its own selfish interests.

However, even before the rupture caused by Deutsche Telekom there were problems with the Global One alliance. Senior staff in the companies told me that there was a significant culture clash between FT and Deutsche Telekom that was negatively affecting their efforts to coordinate their activities and take them forward. This was reflected in the significant delays that occurred when the companies attempted to construct a pan-European network according to commonly agreed standards. Further problems resulted from the fact that each partner used a different equipment supplier—FT tended to use Alcatel while Deutsche Telekom preferred Siemens—and wanted the other partners to use the same equipment it used. Furthermore, each company also had its own private activities in the countries where Global One operated and it is likely that all of them had at least half an eye on the possibility of the alliance falling apart. In the event of this happening they wanted to have at least some fallback position that would not involve starting from scratch in that country.

However, as problematical as the situation was, Global One was able to struggle on. That is until May 1999 when Deutsche Telekom made the dramatic announcement that it proposed to take over Telecom Italia, the incumbent in Italy. The problem was that Ron Sommers, CEO of Deutsche Telekom, had failed to consult, or even inform, his partner and ally in Global One, Michel Bon, the CEO of FT. In retrospect it is clear that Sommers believed that the benefits of acquiring Telecom Italia would outweigh the costs in terms of the damage done to the Global One alliance.

Unfortunately for Sommers, however, things did not go according to plan. In the event it was Olivetti, the Italian business machine and PC company turned telecoms company, that succeeded in acquiring Telecom Italia in a move that shook the Italian business establishment. Forgoing the benefits of acquiring Telecom Italia, Sommers could not avoid the costs of the break-up of Global One. In the event, in January 2000 Deutsche Telekom competed with FT for full ownership of Global One. FT emerged as the winner, paying $4.4 billion for Global One, $3.88 billion of which was paid in cash ($1.13 billion for Sprint's share and $2.75 billion for Deutsche Telekom's) with the remaining amount being paid for the debts that these two partners had accumulated in the alliance company. In fact it is worth noting that the competition for Global One favoured FT. The reason is that its share price had increased relative to Deutsche Telekom's and this made its part of the cross-company shareholding more valuable. Under the agreement Sprint and Deutsche Telekom will continue to give FT access to parts of their international networks for a period of two years, giving the latter time to improve its own international network.

For FT sole control of Global One has boosted its international strategy. For a start, the costs of coordination amongst partners with conflicting interests have now disappeared from Global One. Perhaps FT will also be able to move more decisively into foreign markets. Furthermore, FT will be able to fully integrate its Transpac IP data network, which it contributed to Global One, into its expanding international network. Wasting little time in pushing this expansion forward, FT at the end of 2000 acquired Equant, a global data network that was originally created for the purposes of airline reservations. The only remaining problem with Global One, a problem not to be taken lightly, is one of profitability. Like other global telecoms alliances intended to serve multinational companies—such as Concert that involved AT&T and BT (but broke up at the end of 2001)—Global One is not very profitable. However, Michel Bon is optimistic about the longer term prospects, pinning his hopes on the assumption that for multinational corporations quality and reliability of service is more important than price.

MAJOR ACQUISITIONS AND INVESTMENT

France Telecom's acquisitions and investments reflect its international business strategy outlined in Table 5.8. The company's major acquisitions and investments are shown in Table 5.9.

Table 5.9 shows the wave of acquisitions and investments made by France Telecom from 1996 to 2000. An analysis of Table 5.9 makes it clear how much more focused FT's international strategy has become since the rupture in the company's relationship with Deutsche Telekom. Until May 1999 when this rupture occurred FT did very little on the international front apart from its investment in Global One. It was only from 1999 that France Telecom began to move forward decisively and without the ambivalence caused by the conflicting interests involved in Global One. In the UK, for example, FT made a move into cable TV by investing in 1999 in one of the two major cable companies, NTL, thus providing entry into the broadband access market. Previously, FT's major investment in the UK had been in Metro holdings, a joint venture with Deutsche Telekom and the British new entrant, Energis, in metropolitan area networks.

But it was in 2000 that FT made its major bets in the UK with its acquisition of the second most important mobile operator, Orange, and its acquisition of

Table 5.9. France Telecom: major acquisitions and investments

Date	Acquisitions and investments
1996	Creation of Global One (joint venture with Deutsche Telekom and Sprint)
	Investment in Sprint
	Creation of Wanadoo
1997	Investment in Optimus (mobile, Portugal)
1998	
1999	Investment in NTL (CATV, UK)
	Acquisition of Alapage (e-commerce, France, Europe)
2000	Full acquisition of Global One
	Acquisition of Orange
	Investment in Mobilcom (Germany)
	Acquisition of Freeserve (ISP, UK)
	Share of Wind (Italy) increased from 24.5 to 43.4%
	(the increased shares formerly owned by Deutsche Telekom)

the country's largest internet service provider (ISP), Freeserve. In Germany, FT made its move through its investment in Mobilcom, one of the most aggressive new entrant competitors to Deutsche Telekom. In Italy, FT consolidated its position by increasing its share of Wind, a joint venture with Deutsche Telekom and an Italian partner involved in fixed communications and mobile. In 2000 FT increased its holding in Wind from 24.5 to 43.4 per cent by acquiring Deutsche Telekom's share of this company.

It seems clear, therefore, that if anything the rupture with Deutsche Telekom has been a blessing for FT in so far as it has enabled the company to increase its focus and determination in its international ambitions. This puts another nail in the coffin of the 'Age of Telecoms Alliances' by revealing the extent to which these alliances—based shakily on an ambivalence, if not an outright conflict, of interests—often hindered rather than helped the alliance partners to achieve their international objectives. It is noteworthy that none of the major alliances established during this age have survived.

GROWING IMPORTANCE OF INTERNATIONAL BUSINESS

Table 5.10 reveals just how important international activities are becoming as a source of revenue for FT.

France Telecom has stated that it wishes to keep the contribution of international business to consolidated revenue at around 20 per cent. It is worth noting that this figure is significantly higher than that achieved by AT&T and BT.

INTERNATIONAL BUSINESS ORGANIZATION

In view of the importance of international business for FT, a key question arises as to how these international interests are to be managed. The experience of all

Table 5.10. France Telecom: share of consolidated revenue from international business

Year	Share (%)
1995	2
1998	9
1999	14
2000	20*

Note: *Projected.

Source: France Telecom News, 10 January 2001.

Table 5.11. France Telecom: new organization to drive international business development

New organization's four zones			
Northern Europe and OECD countries	Central Europe	Southern Europe and the Mediterranean	Latin America, Africa, Asia-Pacific
Head: Vice President Bernard Izerable, reporting to FT Group Vice President	Head: Vice President Michel Bertinetto, reporting to FT Group Vice President	Head: Vice President Brigitte Bourgoin, reporting to FT Group Vice President	Head: Vice President Laurent Mialet, reporting to FT Group Vice President
UK, Ireland, Norway, Sweden, Finland, Canada, US, Japan, South Korea, Australia, and New Zealand	Benelux, Germany, Austria, Denmark, Switzerland, Poland, the Baltic countries, Czech Republic, Slovakia, Hungary, and Slovenia	Portugal, Spain, Italy, Yugoslavia, Romania, Moldava, Bulgaria, Albania, Greece, Turkey, Lebanon, Israel, Jordan, Egypt, Tunisia, Algeria, Morocco	(No specific countries mentioned)

Notes:
1. Trans-national services, including Global One, GlobeCast, and the European Backbone Network (EBN) will remain part of their respective operational divisions (i.e. not be part of the above four zones).
2. A strategy unit that will identify international growth opportunities and oversee the consistency of the development plans [of the four zones] will be established in the Development Branch of FT.
3. A new international finance unit (part of the corporate human resources and finance division) has been created to coordinate financing of all international operations.
Source: France Telecom News, 10 January 2001.

multinational companies testifies that this is no easy question. At the beginning of 2001 FT announced an important organizational change aimed at giving the company better focus and control in international markets. The details are summarized in Table 5.11.

Figure 5.3 shows how FT's organization in countries outside France fits in with its organization in France.

As Fig. 5.3 shows, FT's businesses abroad, in those cases where the company has majority control of the business, are organized as separate companies that have two lines of responsibility. (In Fig. 5.3 the UK is taken as an example, although the same principle extends to other countries outside France.) The first line of responsibility is to the unit in FT (France)—see Fig. 5.2 and Table 5.3—that is most closely associated with the overseas business. In the case of the UK, for example, (see Fig. 5.3) Orange is responsible to the Mobile

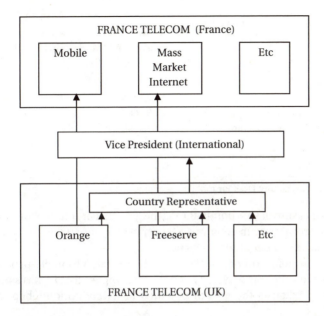

Fig. 5.3. France Telecom: international organization, 2001 (using UK as example).
Source: M. Fransman.

unit (now called New Orange) in France. Similarly, Freeserve, the largest independent ISP in the UK acquired by FT, is responsible to the Mass Market Internet unit (called Wanadoo).

The second line of responsibility is to FT's representative in the country, and, through her or him, to the Vice President (International), also referred to in Table 5.11.

MOBILE

As was shown earlier, mobile communications make an extremely important contribution to FT's revenue and growth. In this section we will analyse how FT is developing its mobile business.

We begin with an overview of the French mobile market, given in Table 5.12.

Several important conclusions emerge from an anlaysis of Table 5.12. The first is the dominance of FT of the French mobile market. This dominance is indicated in two ways in Table 5.12. The one is the company's market share. In 2000 FT accounted for an overwhelming 48.2 per cent of the market. This compares with the second ranked company, Cegetel (including SFR) at 35.6 per cent. Although FT's share is not nearly as large as NTT DoCoMo's (which is near to 60 per cent in Japan), it is significantly more than BT's (now called mmO2) (which is around 30 per cent in the UK compared to Vodafone's 37 per cent).

Table 5.12. French mobile market, 2000

Company	Number of subscribers	Market share (%)	Market growth (%)
France Telecom (including Itineris, Ola, and Ameris)	10,926,400	48.2	44.9
Cegetel (including SFR, GSM, and SRR)	8,069,600	35.6	33.6
Bouygues (Bouygues Telecom)	3,649,600	16.1	21.5
Total	22,645,600		
Penetration rate	37.7%		

Source: ART, Observatory Desk for mobile telephony, 2000.

The other measure of FT's dominance is the growth in its number of subscribers. For FT the growth rate in 2000 was 44.9 per cent, compared to 33.6 per cent for Cegetel and 21.5 per cent for Bouygues.

The second point to note is the relatively low rate of mobile penetration in France that in 2000 was only 37.7 per cent. This implies that FT has some way to go before it exhausts the growth potential of its domestic mobile telephony market.

France Telecom's positioning in the European mobile market took a great leap forward in 2000—part of the benefit of the company's increased focus on the international arena—when it acquired 100 per cent of the second most important British mobile operator, Orange. (Orange was owned by Hutchison of Hong Kong when it was audaciously acquired by Mannesmann. This invasion of the UK market by the German company provoked Vodafone, the UK mobile market leader, into an eventually successful hostile acquisition of Mannesmann itself. However, one of the requirements of the regulators was that Vodafone sell-off Orange in order to avoid undue concentration particularly in the UK market. France Telecom became the successful buyer of Orange.)

Orange was particularly attractive for FT. With a reputation for dynamic marketing and with the fastest growth rate in the UK mobile market Orange provided rich pickings. Moreover, it had won a third-generation UMTS mobile licence in an auction from which France Telecom had withdrawn due to the high price. Although FT paid a price for Orange that some financial analysts thought too high, it left FT in a very strong position in the European mobile market as a whole (benefiting from Orange's other European investments). Indeed, the acquisition made FT the second largest mobile player in Europe. The market shares of the major European mobile players are shown in Table 5.13.

France Telecom decided to merge Orange with its other mobile interests, including Itineris in France, and call the new company New Orange. In February 2001 FT floated 15 per cent of New Orange's shares, partly in order to raise the funds to reduce its significant level of debt to which high third-generation mobile licence fees contributed.

Table 5.13. Mobile market shares in the European market

Company	European market share (%)
Vodafone	30
FT/New Orange	25
Telecom Italia	15
Deutsche Telekom	10
BT	8

Source: J. Krafft, *France Telecom: From a Phone to a Net Company*, CNRS-IDEFI-LATAPSES, December 2000 (mimeo).

Third-generation licences were also a significant factor motivating FT's investment in the German telecoms company, Mobilcom. In March 2000 FT acquired 28.5 per cent of Mobilcom and went on in August 2000 to acquire with Mobilcom Multimedia a third-generation UMTS mobile licence for Germany. In September 2000 FT acquired a UMTS licence in the Netherlands through the company Dutchstone.

To conclude, with its acquisition of Orange and the formation of New Orange, FT is well positioned to take advantage in Europe of the promises of third-generation mobile demand. How extensive and profitable this demand will be remains to be seen, an uncertainty that faces not only FT but also the leader, Vodafone, and the other mobile operators in Europe and around the world.

INTERNET

Unlike BT which dilly-dallied before making a firm commitment to the Internet, FT moved more decisively into this market as early as 1996 when it established its Internet subsidiary, Wanadoo. This move was made despite the slow uptake of the Internet in France, partly as a result of the success of France's earlier home-grown (though originally Internet-incompatible) online information service, Minitel. Realizing the importance of the Internet, in February 1998 FT entered into an agreement with MSN (Microsoft Network) whereby it acquired the latter's customers and was able to provide a range of Internet services in addition to simple Internet access. In March 1998 FT acquired 67 per cent of Oleane, a pioneer of Internet access in France, and in July 1998 it acquired Oda, an online directory services.

Table 5.14. France Telecom/Wanadoo: breakdown of revenue

Activity	1999 revenue (% of net revenue)	2000 revenue (until 30 September 2000, % of net revenue)
Directory	86.0	70.6
Internet access	12.0	23.6
Portals	1.5	3.4
E-commerce	0.5	1.8
Services for enterprises	~0	0.6
Other services	0.4	0.1
Of which international	32.4	40.9

Source: Wanadoo.

In Table 5.14 a breakdown is given of Wanadoo's revenue.

Table 5.14 provides an interesting insight into the sources of revenue of Wanadoo while at the same time pointing to some of the difficulties facing Internet companies in general. Significantly, almost 95 per cent of Wanadoo's revenues came from only two sources. The first source, directory services, accounted for the large majority of the company's revenue, namely 70.6 per cent. The second source was Internet access that provided 23.6 per cent of revenue. Portals provided only 3.4 per cent of revenue while the much-hyped e-commerce only generated 1.8 per cent of revenue. Notably, international Internet services (i.e. outside France) provided 40.9 per cent of revenue, high-lighting not only Wanadoo's international acquisitions but also its position in the world of the French-language Internet.

The dilemma posed for Wanadoo and also for its competitors is clear. While Internet access is an important source of revenue (growing in proportional significance between 1999 and 2000) it has increasingly become a 'commodity' that can be provided by many other companies. Although the barriers to entry in the provision of directory services are much higher, Wanadoo's advantage in this area is more a reflection of FT's legacy as a monopoly provider of telephone directories and yellow pages than anything else. However, 'New Age' Internet-based services such as portal services and e-commerce, for the moment at any rate, are not significant revenue spinners.

In July 2000 FT floated 10 per cent of Wanadoo and in September 2000 Wanadoo acquired Freeserve, the UK's largest ISP established by the retail electronics company, Dixons. This confirmed Wanadoo as one of the leaders in the rapidly consolidating European ISP market. The other leaders include AOL and Yahoo! from the US and T-Online (Deutsche Telekom), Terra Networks (Telefonica, that acquired the American ISP, Lycos), Tiscali (Italy), and World Online (the Netherlands) from Europe.

It may be concluded, therefore, that although FT has moved quickly, ener-getically, and successfully into the Internet age, it is not yet seeing significant earnings and profit benefits in this area. This conclusion is evident also in Table 5.5 (where Internet-related activities in France are included largely in 'leased lines and data') and in Table 5.6 (where they are included in 'other services in France'). This reinforces our earlier conclusion that for FT it is activities in mobile and international, rather than data and Internet, that are accounting for the company's current dynamism and relatively strong performance. We shall return to this important point in the concluding sections of this chapter.

R&D

Before FT's privatization the company's main central R&D laboratories was CNET. Like NTT's Electrical Communications Laboratories, CNET, the largest telecoms R&D organization in Europe, was regarded by the government as having public research obligations. CNET developed a distinguished research record, making important contributions in areas that included digital switching (the Plato digital switch developed by CNET has a claim to being one of the first digital switches developed in the world), Minitel (France's pre-Internet online information service), and ATM (asynchronous transfer mode) switching.

With the privatization of FT, however, CNET was abolished and replaced by FT R&D. Currently, FT R&D employs 3,800 staff of whom 3,000 are research engineers. Some 80 per cent of the products and services commercialized by FT have been developed by FT R&D which owns around 4,000 patents. FT R&D is based in eight different sites in France and has one site in Brisbane, Silicon Valley, California. The latter site focuses on three areas: technology watch, system integration, and software.

With the abolishment of CNET, FT R&D was encouraged to focus its research far more narrowly on FT's needs although, as will shortly be seen, again like NTT's Electrical Communications Laboratories, FT R&D has retained some of its public research obligations. The tighter focus of FT R&D on the needs of FT together with the abandonment of several costly areas of research to specialist technology suppliers—such as switches and electronic and optical devices— has significantly reduced FT's R&D intensity. This is shown in Table 5.15.

As Table 5.15 clearly shows, FT's R&D intensity has been on a steady downward trend. The decrease is far more rapid than three other of the Big Five incumbents (BT, Deutsche Telekom, and NTT). Only AT&T's R&D intensity has decreased more rapidly, a reflection of the company's divestiture of Lucent in September 1995. The decrease in R&D intensity in FT is primarily due to the increasing degree of vertical specialization in the Telecoms Industry as a whole which has meant that a greater proportion of telecoms technologies are now supplied by specialist technology suppliers rather than the central R&D laboratories of the incumbents as was the case before liberalization and privatization (analysed in Chapters 1 and 2).

Table 5.15. France Telecom: R&D intensity

Year	R&D ratio (R&D expenditure as % of consolidated revenue)
1995	3.7
1996	3.6
1997	3.4
1998	2.7
1999	2.2
2000	1.3

Source: France Telecom Annual Reports and Form F20, *Financial Times*, R&D Scoreboard, 27 September 2001.

Table 5.16. France Telecom: public R&D obligations and R&D strategy

Public R&D obligations	R&D strategy
'Pursuant to the Public-Interest Service Charter, France Telecom SA must spend not less than 4% of its gross revenues (excluding revenues from its subsidiaries) on research and development. This level of spending (expenses and capital expenditures) is in line with France Telecom's recent history and the Company considers it to be consistent with its future research and development plans.' (p. 55)	'France Telecom's research and development priority is the development of new services and network architectures and its activities have been oriented toward high capacity transmissions, the Internet and multimedia, as well as intelligent networks.' (p. 55)

Source: France Telecom, Annual Report, 1999.

However, as already noted, FT R&D has retained public service research obligations. This feature is unique to the French and Japanese national systems. The obligations of FT R&D are summarized in Table 5.16.

Several points emerge from Table 5.16. The first is the obligation that the French government has imposed on FT, in the form of a Public-Interest Service Charter, to spend not less than 4 per cent of its gross revenue (excluding revenue from its subsidiaries) on R&D. It is important to note that this public-interest obligation is significantly different from that which the Japanese government has imposed on NTT. More specifically, in the Japanese case, although the amount of R&D spending required is not specified, NTT is required to give outside access to the fruits of some of its R&D. The ambiguity of this requirement has long posed a problem for NTT. The reason is that it is not clear what R&D NTT is supposed to provide outsiders access to and, furthermore, it is unclear what 'providing access' means. The fact that simultaneously NTT is required to compete with new competitors further complicates this requirement.

It may be concluded, therefore, that although both FT and NTT have public-service R&D obligations, in NTT's case the obligations are ambiguous and cause more problems for the company.

The second point emerging from Table 5.16 is that FT appears to be able to live comfortably with this R&D obligation. The reason is that the requirement is well within the boundaries of R&D expenditure that the company would in any event carry out. Furthermore, there is no requirement for FT to make its R&D output available to competitors and other national outsiders, unlike in NTT's case.

The third important point contained in Table 5.16 is further information on FT's R&D strategy. According to this strategy, the company will concentrate on the following areas for its R&D:

- development of new services
- development of new network architectures
- high capacity transmissions systems
- internet and multimedia and intelligent networks.

Notably absent from this list of strategic priorities for FT's R&D are areas such as switching and electronic and optical devices, areas that previously consumed a significant amount of the company's R&D budget. As noted earlier, this new list of R&D priorities is also a reflection of the advancing process of vertical specialization in the Telecoms Industry as a whole which has meant that areas like switching and devices can now be left by network operators like FT to specialist equipment and technology suppliers. Also implicit is the fact that the five areas identified in this list of R&D priorities are areas that, while important for FT's growth and competitiveness, cannot as satisfactorily be catered for by specialist equipment and technology suppliers.

Innovation in R&D

Innovacom and France Telecom Technology. One of the innovations introduced by FT and FT R&D has been measures to encourage spin-off companies as a way of commercializing R&D. Such spin-offs are assisted by the establishment by FT of an internal venture capital company, Innovacom. The spin-off programme in FT is still new and the number of spin-offs is relatively small. However, one highly successful example is Algety Telecom, established from research done by FT R&D, which produces soliton-based optical transmission systems. Recently Algety was acquired by the US optical fibre equipment company, Corvis. Algety also received assistance from FT Technology, a new organization established by FT R&D to provide assistance to spin-offs in the form of know-how, software, and patents.

FRANCE TELECOM AND THE OTHER
BIG FIVE INCUMBENTS

As the present chapter makes clear, FT has decided to remain as a loosely integrated company. In doing so it has come to the same conclusion as Deutsche Telekom (discussed in Chapter 4) and NTT (in Chapter 3). However, AT&T has chosen to break up into four separate companies (Chapter 3 and Appendix) while BT has chosen to separate only its wireless subsidiary (Chapter 3).

THE FUTURE OF FRANCE TELECOM

There are two very different stories that can be constructed regarding FT's future. While the first is optimistic, the second is pessimistic.

The Optimistic Story

According to the optimistic story, even though FT only faced full competition from 1 January 1998, the advent of competition will not seriously erode the company's future revenue growth and earnings. The reason is that, unlike AT&T and BT, FT has been able to dominate the major markets in its home country, France. Perhaps because it has been able to learn from the experiences of AT&T and BT who faced competition from the mid-1980s, FT moved quickly and effectively to establish a commanding position in the growth areas of mobile, Internet, data, and international activities in France.

This is certainly the picture that emerges from Fig. 5.2 (that shows FT's market share in France) and Table 5.6 (that shows its business performance by business segment). In contrast, AT&T was hamstrung by legislation that separated it from local access and left it vulnerable to the commoditization of long-distance telephony while BT focused more on its ill-fated attempt to cement an alliance in the US than on its local markets.

But what about the threat to FT of increasing competition and commoditization in domestic markets? For example, will not local loop unbundling (regulations that require FT to allow competitors into its local networks to provide competing services) rapidly erode FT's revenue and earnings?

It is true that some French telecoms markets are like the US long-distance voice market that has so troubled AT&T's Michael Armstrong. In these markets there are a large number of entrants, they face few technical constraints (as a result of the help provided by specialist technology suppliers like Alcatel, Nortel, Lucent, and Ericsson), and there is little to differentiate the services provided by the different competitors (i.e. there is commoditization). In markets such as these prices fall and margins are sashimi-thin. Only large players, riding on economies of scale, can survive.

However, while these markets are important for FT, many other of the markets in which FT competes are quite different. Take, for example, the key services that are provided to businesses, large and small. While every business does need a good voice service, equally important are other services such as Internet access, web site design, maintenance, and hosting, and mobile services that soon will include various forms of communication with customers and suppliers through mobile phones and computers. Already these kinds of services are becoming important for even small businesses. And large businesses need even more sophisticated forms of e-interaction. As the dominant diversified telecoms services provider, FT is in a strong position to offer the full range of services to its business and residential customers, an offering that cannot be matched by its more narrowly-focused new entrant competitors.

Furthermore, as a result of factors such as economies of scale and the Telecoms Bust, FT may be one of the major beneficiaries from the processes of shakeout and consolidation that result in oligopolistic market structures. In these kinds of markets FT's size, reputation, brand name and wide range of services may be used to advantage.

To the extent that the optimistic story comes to pass, the future for FT, like its Orange mobile subsidiary, will be bright. France Telecom will still have to face its many competitors building pan-European networks—numbering some fifteen companies according to a recent count, including aggressive US entrants, such as WorldCom, Qwest, Global Crossing, and Level 3. However, these rivals will not have the 'breadth' to provide the full range of services that FT provides. In such a world FT's future looks relatively secure, as does the future of most telecoms incumbents, provided they are able to dominate their domestic markets.

The Pessimistic Story

However, those wanting to construct a pessimistic story about FT's future will challenge the core assumptions of the optimistic story. They will argue that FT and the dominant incumbents in other national markets are vulnerable to a two-pronged attack. The first prong is precisely that which is acknowledged in the optimistic story, namely the prong of commoditization. The devastation that this prong can cause is most dramatically evident in the case of AT&T and its falling revenue from long-distance telephony. BT has also suffered from precisely the same attack, although, unlike AT&T, it has received some cushioning from its strength in local access which has not yet been undermined by a combination of competing local access technologies and regulation.

The second prong, that together with the first could form a pincer-squeeze, is the vulnerability of even specialized and complex service markets—such as e-transactions, e-communications, broadband access, and mobile access—to attack by narrowly-focused, specialized competitors. According to the pessimistic story these markets are vulnerable to highly focused new entrants who are capable of rapidly providing competitive and innovative services. The reason is that these new entrants are able to take advantage of the latest technologies provided by specialist technology suppliers and are able to focus on narrowly targeted market segments. Being tightly focused, and able to buy all the latest technology that they need on competitive and well-functioning technology markets, they can remain far smaller than widely diversified incumbents like FT. Their F4 characteristics—focused, fast, flexible, and flat—will give them a competitive advantage.

CONCLUSIONS

In this chapter it has been shown that FT has managed so far to dominate its main domestic markets. Furthermore, it has been able to benefit from Europeanization that has provided an important additional source of growth. The abandonment of FT's alliance with Deutsche Telekom and Sprint, Global One, has helped the company to further its European ambitions. An important downside, however, one that troubled the company in the first years of the new millennium, was the high level of indebtedness that resulted from expensive acquisitions and (to a lesser extent) the high cost of third-generation mobile licences.

Two stories were constructed in order to analyse what will happen to FT in the future. Which of these stories will come true? Only time will answer this final question. Under the conditions of interpretive ambiguity that surround both stories, optimists and pessimists will continue to contend.

6

The New Entrepreneurs in the Telecoms Industry

In Chapters 1 and 2 the importance of vertical specialization and its consequences are analysed. The consequences include lowered technological barriers to entry, rapid entry by aggressive new entrants, and over time falling rates of revenue growth and profitability resulting from the increased competition following new entry.

In this chapter attention is focused on several of the most important new entrant network operators that emerged in the US in the 1980s and 1990s to compete with the incumbent, AT&T. Not only did these new entrants grow rapidly in the US, they also quickly globalized their operations.

Five questions are addressed regarding these new entrants. First, from what backgrounds did they enter the Telecoms Industry? Second, how much prior knowledge did they have of this industry? Third, from where did they get the initial capital that was necessary to enter the industry? Fourth, what does the experience of these entrepreneurs and their new entrant companies teach about the dynamics of the Telecoms Industry in the Internet Age?[1] Fifth, what happened to them as a result of the Telecoms Bust?

INTRODUCTION

In early 2000 the state of the global Telecoms Industry could not have seemed rosier. Telecoms share prices were at an all time high. Optimistic expectations

The author would like to acknowledge with gratitude excellent research assistance in the preparation of this chapter from Ian Duff.

[1] The focus of attention in this chapter is on the entry process, rather than on the subsequent performance of the new entrants. While, as is shown in Chapter 1, the new entrants in the US and UK tended to perform significantly better than the incumbents from around 1996 to about March 2000 in terms of stock market value, subsequently their relative performance declined substantially. Indeed, by the beginning of 2002 two of the five companies studied in this chapter had become bankrupt. The reasons for this are analysed in Chapter 1.

were fuelled by exploding demand for Internet access and applications and by rapid growth in mobile communications. These expectations fed into the already buoyant tide of the bull stock markets that had been sweeping North America and Europe for an unprecedented period. So heady were the times that some economists and policy makers were even able to persuade themselves that a 'new economy' had dawned, with rules fundamentally different from those of the old.

But as the year drew to a close the picture could not have been more different. The share price of many telecoms companies, previously regarded as the pick of the rosy bunch, had fallen more than 50 per cent below the year's highs. In AT&T's case, for example, the share price was lower than when CEO Michael Armstrong took over in 1997. BT too had hit the doldrums with an equally spectacular slump in its share price. Both these companies, previously the lions of their jungles, were forced by irate shareholders and by internal attempts to 'get back on track' to institute drastic corporate reorganizations. In AT&T's case the astounding decision was made to breakup the grand company, established in 1885, into four separate companies. But it is not only the incumbents, the beneficiaries of previous years of regulated monopoly, who were hit. WorldCom, the most successful of all the new telecoms entrants in the world, was equally punished.[2] By the beginning of 2002, however, the financial situation of all the new entrants had deteriorated even further. Indeed, two of the five companies studied in this chapter had declared bankruptcy.

Why such a dramatic change in such a short period of time? While the causes are multiple and complex (and are examined in detail in Chapter 1) here attention will focus only on one important contributory factor. This is the aggressive new entrants that rapidly emerged after countries and regulators made the decision to liberalize their telecoms services industries and allow competition. Specifically, this chapter focuses attention on some of the most aggressive of these new entrants in the US, which quickly went on from their American roots to tackle global markets. Here the spotlight will be put on the entrepreneurs that were the prime movers.

THE ENTREPRENEURS AND THEIR COMPANIES

The entrepreneurs, their companies, and the founding date are shown in Table 6.1.

As shown in Table 6.1, five US new entrant telecoms network operators have been selected for study. They were among the most successful of the new entrants. On 29 November 2000, that is some eight months after the 'telecoms crash' began in March 2000, they had a combined market capitalization of

---

[2] These events are analysed in Chapters 1 and 3.

Table 6.1. The entrepreneurs and their companies

Company	Entrepreneur	Founding date	Market capitalization ($ bn, 29 November 2000)
Global Crossing	Gary Winnick	1997	12.6
GTS	George Soros	1983	0.3
Level 3	Walter Scott Jr and Jim Crowe	1997	11.5
Qwest	Philip Anschutz	1988	68.1
WorldCom	Bernard Ebbers	1983	44.4

$136.9 billion. Three of the companies—GTS, Qwest, and WorldCom—were started in the 1980s, two in the early 1980s, and one in the late 1980s. Two of the companies—Global Crossing and Level 3—were started late in the 1990s.

STOCK MARKET PERFORMANCE—1999 AND 2000

How well have these five companies performed? In Tables 6.2 and 6.3 performance is measured according to stock market capitalization. In both these tables the five companies are compared to a number of other companies, mainly also in the telecoms sector. The dates of comparison are July 1999, before the telecoms crash, and November 2000, eight months after the crash.

A number of interesting points emerge from Table 6.2. The first is the significant total market capitalization of the five companies (shown in bold in the table), totalling $235 billion in July 1999. (A year and a half later, in November 2000, eight months after the crash, $100 billion would be wiped off their combined value.) The second point is the remarkable achievement of WorldCom, then the fourteenth most valuable company in the world. With a market capitalization of $152 billion, WorldCom was not far from AT&T, then the most valuable telecoms network operator, the seventh most valuable company, with a market capitalization of $186 billion. Indeed, WorldCom's value was almost the same as NTT's, at $157 billion, the thirteenth most valuable company.

The third point is the astounding progress made by the two companies that had only been launched in 1997. Global Crossing, which began in March 1997 had a market capitalization of $20 billion by July 1999. Level 3's value at the same date was $27 billion, a company that began in June 1997.

By the end of 2000 the fortunes of individual companies had changed somewhat, as is shown in Table 6.3 (that ranks only the fifteen companies in this sample).

There are several notable points to be made about Table 6.3. The first is the reversal in fortune for several of the individual companies. Particularly noticeable is the fall of WorldCom, from a market capitalization in July 1999 of

Table 6.2. Market capitalization of some telecoms new entrants and other selected companies, 12 July 1999

Name	Rank	Market value ($ bn)	Country
Microsoft	1	407	US
AT&T	7	186	US
Cisco	9	174	US
NTT	13	157	Japan
MCI-Worldcom	14	152	US
Lucent	16	150	US
Deutsche Telekom	23	115	Germany
BT	26	107	Britain
NTT DoCoMo	27	106	Japan
SBC Communications	31	100	US
FT	43	80	France
Telecom Italia	58	67	Italy
Nortel Networks	84	50	Canada
Qwest	146	30	US
Level 3	172	27	US
Williams	204	22	US
Global Crossing	244	20	US
NTT Data	255	19	Japan
GTS	716	6	US

Source: *Business Week*, 12 July 1999.

Table 6.3. Market capitalization of selected incumbents and new entrants, 29 November 2000

Company	Symbol	Market capitalization ($ bn)	Rank
Vodafone	VOD	208.5	1
NTT	NTT	132.4	2
Deutsche Telekom	DT	98.5	3
FT	FTE	90.0	4
Telecom Italia	TI	85.1	5
AT&T	T	70.1	6
Qwest	Q	68.1	7
Telefonica	TEF	63.3	8
BT	BTY	57.5	9
WorldCom	WCOM	44.4	10
COLT	COLT	13.0	11
Global Crossing	GX	12.6	12
Level 3	LVLT	11.5	13
Energis	ENGSY	10.4	14
GTS	GTS	0.3	15

$152 billion to $44 billion by the end of November 2000. Equally noticeable is the corresponding rise of Qwest through the ranks. Its capitalization rose from $30 billion to $68 billion, a remarkable achievement at a time of slump in telecoms share prices. The collapse of GTS is also noteworthy, from $6 billion to $0.3 billion.

The reason for WorldCom's fall is much the same as AT&T's and BT's (analysed in Chapter 3). In short, all three companies were hit by falling revenue and profit growth from fixed voice services. Although all companies experienced better performance in the 'new growth areas' of Internet, data, and mobile, this performance was insufficient to plug the gap in fixed voice. GTS's collapse was largely due to worries that there would be a glut of pan-European network capacity and to the crisis in the Russian economy since 1998. Ironically, the problems affecting all these companies stemmed in large measure from the effects of the increase in the intensity of competition caused by the new entrants collectively. This resulted in falling prices and therefore a fall in revenue and profit growth. By November 2000 the market capitalization of both Global Crossing and Level 3 had fallen, with the latter taking a harder knock. But their fall was not as great as that of AT&T and BT.

Qwest, however, was able to buck the trend for reasons which include the company's successful move into local services with its acquisition of the Baby Bell, US West, its move into higher-margin higher layers of the Telecoms Industry in areas such as web-hosting and application service provision, and its successful global moves such as KPN-Qwest, its joint venture in Europe with the Dutch incumbent.

Two other notable points emerge from Table 6.3. The first is that the new entrants, taken as a whole, had performed respectably. As already noted, the five new entrants in November 2000 had a combined market capitalization of $137 billion, slightly more than that of NTT at $132 billion. And the five companies, it must be recalled, were only relatively recent start-ups.

Second, however, the incumbents as a whole were alive and well and, in November 2000, also seemed to be doing fairly well. Here the success stories were France Telecom (FT) (analysed in Chapter 5) and Telecom Italia, whose market capitalization had increased over the period, although NTT and Deutsche Telekom (see Chapter 4) recorded only relatively slight decreases. It was AT&T and BT that took the most severe knock (for reasons that are analysed in Chapter 3 and the Appendix). AT&T's market capitalization fell from $186 billion to $70 billion, while BT's fell from $107 billion to $58 billion.

Finally, the outstanding success story of the lot was Vodafone, a company that did not even appear near the top of the charts in July 1999. Its success was due to its single-play focus—unfettered by falling revenue growth and profitability in core long-distance and international voice services—and its remarkable growth by acquisition from 1998.

We turn now to examine the questions referred to earlier regarding the five new entrants and their founding entrepreneurs.

THE ENTREPRENEURS

Gary Winnick—Global Crossing

Gary Winnick made his reputation at the firm Drexel Burnham Lambert where during the 1980s he worked with the 'junk bond king', Michael Milken. It was here that Winnick was involved in the financing of MCI's start-up activities (MCI with Sprint being the first two long-distance competitors to AT&T after the latter's divestiture).

In 1985, Winnick left Drexel to form his own investment partnership, Pacific Capital Group (PCG). To begin with, Winnick maintained strong ties with Drexel, but these were loosened after the court case against Milken. Although Winnick was not himself charged, he was ordered by a court to testify in Milken's case in return for immunity. In the event, being charged with having violated securities laws, Milken reached agreement with the prosecutors before testimony took place.

Winnick's investment performance with PCG was unremarkable. In late 1996, Winnick encouraged a PCG partner, David Lee, to speak to people in AT&T in order to look for investment opportunities. AT&T proposed that if Winnick could raise $750 million it would guarantee the construction of an undersea telecoms cable linking the US and Europe. AT&T's rationale, apparently, was that it wanted to spin-off its cable-laying operation and thought that this would be facilitated by a large existing contract.

Winnick raised the money, contributing $15 million of his own. At this stage he had very little prior knowledge of the Telecoms Industry and had to watch a video in order to learn more about the industry in which he was beginning to invest heavily.

In the beginning Winnick found it difficult to sell capacity on the undersea cable still under construction. However, things changed in October 1997 when he convinced a number of telecoms executives to buy capacity by slashing the price. At the time the going rate for a 155 megabit per second circuit was around $20 million. Winnick, benefiting from the latest technologies, reduced the price to $8 million. These deals allowed him to allocate only 10 per cent of the cable's capacity, but to virtually recoup almost half of the total investment in the cable. This reflected both the scarcity of capacity that existed at the time as well as the falling costs made possible by the latest generation of technologies. At this stage, with only one cable, the company was called Atlantic Crossing. Later, with plans to repeat the trick, the company's name was changed to Global Crossing.

Aware that he was onto a good thing, Winnick used the company's shares in order to recruit the kind of people he needed with the necessary skills. These included Robert Annunziata who was head-hunted from AT&T to become the new CEO of Global Crossing in February 1999. Annunziata had left AT&T earlier to establish Teleport, a competitive local exchange carrier (CLEC) that was established by Fidelity, the largest mutual fund in the US, and Merrill Lynch. Teleport started its activities in New York and then replicated its business plan in the Boston area. In July 1998 AT&T acquired Teleport in a deal worth $11.3 billion. With this deal Annunziata returned temporarily to AT&T until he was head-hunted for Global Crossing.[3] However, in little over a year Annunziata left Global Crossing, when he was replaced in March 2000 as CEO by Leo Hindery, formerly the CEO of the cable television company, TCI, that was acquired by AT&T.

By the end of the 1999 financial year, Winnick owned 12 per cent of the shares of Global Crossing. Furthermore, the PCG, the investment company founded by Winnick, also owned 12 per cent. The directors and other executives of Global Crossing, including Winnick, owned 25 per cent.[4] According to *Forbes*, Winnick would go down in history as the man who made the fastest fortune, ever.[5]

George Soros—Global Telesystems (GTS)

The history of GTS goes back to 1983. In this year an organization, called San Francisco/Moscow Teleport, was established as a non-profit venture to improve communications between Russian and US universities. The venture was funded by a philanthropic trust established by the Hungarian-born investor, George Soros. Soros made his fortune largely through the Quantum Investment Fund which contributed to his alleged personal fortune of over $5 billion in 1997. Committed to the transformation of societies in accordance with so-called 'open society' principles, Soros established the Open Society Foundation to further these aims. Russia and Eastern Europe became a priority for this Foundation.

Having established its network in Russia, the venture soon realized that the backward state of the country's telecommunications system created important commercial opportunities. In 1986 the venture was incorporated as a for-profit company. In February 1995 the company's name was changed to Global Tele-systems Group, Inc. In 1995 GTS established a joint venture with eleven European railway companies, called Hermes Raitel (HER), the aim of which was

---

[3] In Boston, Teleport employed Paul Chisholm, who originally trained as an economist. Fidelity moved Chisholm to the UK in 1992 when the company replicated its business plan in that country with the establishment of City of London Telecommunications (COLT). COLT, referred to in Table 6.3, became one of the best-performing companies in the world, measured in terms of the appreciation of its share value, and Chisholm became the best-paid executive in the UK until his retirement from the company early in 2000. COLT is discussed in detail in Chapter 7. [4] Global Crossing, Annual Report and Form 10K, 1999.

[5] *Forbes*, 19 April 1999.

to set up a high-speed telecoms network across national boundaries in Western Europe. By the end of 1998, GTS owned 86 per cent of HER. In December 1998 GTS acquired another successful pan-European network operating company, Esprit, in a deal worth $757 million. In April 1999 GTS bought 52 per cent of Omnicom (France) that provided services to small and medium sized companies in France, for $210 million.

As one of GTS's original founders, Soros continued to hold a significant minority stake in the company. In 1997 his stake was 26 per cent. Another co-founder with Soros was Alan Slifka whose Abraham Fund is geared towards international reconciliation (particularly between Jews and Arabs). He also had an important minority holding in GTS.

Walter Scott Jr and James Crowe—Level 3

The history of Level 3 goes back to a construction company, Peter Kiewit Sons' Inc (PKS), located in Omaha, Nebraska, that diversified into mining, information services, and communications. In 1985 PKS established a wholly-owned subsidiary, Kiewit Diversified Group Inc (KDG), to hold its non-construction assets. At this time the Chairman of PKS was Walter Scott Jr.

James Q. Crowe, the son of a much-decorated Second World War marine, joined PKS in 1986. After studying as a mechanical engineer at Rensselaer Polytechnic, New York, Crowe completed an MBA at Pepperdine University in California. He then found employment in the construction of electric power plants for Morrison Knudsen. When he was denied promotion in 1986 he left to join PKS.

In PKS Crowe and Walter Scott studied the new opportunities that were being created by electricity deregulation in the US. This led them on to examine the business possibilities being created by the deregulation of telecommunications in the aftermath of the divestiture of AT&T. In turn, this resulted in the establishment of a telecoms network company, Metropolitan Fiber Systems Communications (MFS). The aim of this company was to construct optical fibre metropolitan networks in major US cities, and the company's activities were soon replicated in Europe. With KDG having invested $500 million in MFS, the latter went public in 1993 with Crowe the CEO.

In Omaha, Nebraska, PKS's headquarters was in the same building (known as Kiewit Plaza) as Berkshire Hathaway, the company established by the famous investor, Warren Buffett. Indeed, Walter Scott had been a long-serving Director of Buffett's company. In 1995 Scott attended a meeting that Buffett arranged in Dublin. At this meeting Bill Gates, newly converted as a believer in the Internet, gave a talk on the future potential of the Internet. This talk convinced Scott that MFS should integrate the Internet into its strategy. Accordingly, in April 1996 MFS paid $2 billion for UUNet, the dominant Internet service provider (ISP) in the US. (Microsoft owned 14.7 per cent of UUNet and did extremely well out of the 'stunning' price paid for UUNet.)

In 1996 MFS (including UUNet) was acquired by WorldCom for $14.3 billion. This deal gave WorldCom, until then largely a long-distance company, local access in major US cities, an important Internet backbone, and a foot in the European door. Crowe joined WorldCom after the takeover.

Having sold MFS, however, Walter Scott remained convinced in the profit potential in telecoms and the Internet. It was out of this conviction that Level 3 was born. On 19 June 1997, about a year after MFS was sold to WorldCom, another telecoms company was established by KDG with Jim Crowe, enticed back from WorldCom, as President and CEO. In rejoining Scott, Crowe was able, much to the annoyance of WorldCom's CEO Bernard Ebbers, to take with him many of the old MFS executives who had also joined WorldCom.

On 31 March 1998 it was announced that the company, Level 3 Communications Inc., would be separated from the rest of KDG. (The name Level 3 came from the layered set of protocols of which level 3, associated with routing switches, was the most sophisticated.) Remaining in Level 3 was PKS Information Services—involved in outsourcing, systems integration, and Internet solutions— and the coal-mining subsidiary, KCP. These were intended to give the fledgling company a financial boost. Level 3's headquarters was moved to Denver, Colorado, which was also host to Qwest and the Baby Bell US West (which was acquired by Qwest).

The inspiration for Level 3 was to create an end-to-end Internet Protocol (IP) network and over it to provide Internet-related services, including web-hosting and application services. The belief driving this inspiration was that IP technology created the possibility of significantly cheaper services relative to conventional telecoms services over circuit-switched networks, a belief supported by leading telecoms analysts, such as Salomon Smith Barney's Jack Grubman. According to Forbes' list of the 400 richest people in the US, Walter Scott was ranked in 55th place, with a fortune estimated at $3.8 billion in 1999.[6]

Philip Anschutz—Qwest

Qwest owes its birth to the entrepreneurial energies, property rights, and accumulated capital of Philip Anschutz who, by the late 1990s, was thought to be one of the ten richest people in the US.[7] Anschutz began his business life in the 1960s in the area of cattle ranching and was fortunate enough to discover oil on his property which enabled him to sell his oil-rich ranch for the sum of $500 million. In 1982 he bought the Denver and Rio Grande Western railroad company for $500 million, $90 million of which was paid in cash. In 1986 he

---

[6] Forbes web site, 10/9/99.   [7] *Financial Times*, 30 October 1998.

leveraged this investment to purchase the Southern Pacific Railroad Corporation, a company that he sold in 1995 to Union Pacific for $3.9 billion, bringing him a profit of about $1.6 billion.[8]

It is here that property rights enter the Qwest story, more specifically the rights of way that allow Qwest to bury its conduits and optical fibre cables alongside railway tracks. For in selling the Southern Pacific Railroad to Union Pacific in 1995 Anschutz acquired the right to lay optical fibre cables along the tracks of both companies. In so doing, Anschutz was pursuing a strategy that he had chosen in 1992 when he established an agreement for Qwest, which he had set up in 1988, to acquire rights of way from the Southern Pacific Railroad Corporation. Qwest's rights of way were further enhanced when Anschutz acquired similar rights from the CSX Corporation in 1996 pertaining to 19,000 miles of track.[9]

It was the capital that Anschutz accumulated in cattle ranching, oil, and railroads that enabled him to invest in Qwest. It has been suggested that Anschutz's initial investment in Qwest amounted to $55 million.[10] Qwest was wholly-owned by Anschutz Company until 27 June 1997 when the company issued common stock in its initial public offering. From then Anschutz's personal holding in the company was slowly diluted. On 31 December 1998 Anschutz owned approximately 46.2 per cent of the outstanding common stock of the company.[11] By July 1999, after Qwest's acquisition of US West was announced, Anschutz owned about 39 per cent of the outstanding shares of the company,[12] remaining the largest single shareholder.

While Anschutz's personal capital provided the initial source of investment in Qwest, capital markets played a key role in multiplying the resources at the disposal of the company. Capital markets played this role in two distinct ways. First, by making finance directly available to the company and, second, by significantly and rapidly increasing the value of Qwest's shares that in turn gave the company additional purchasing power with which to acquire further assets through merger and acquisition using its 'paper' as currency.

Bernard Ebbers—WorldCom

MCI-WorldCom proudly traces its origins to a meeting held in September 1983 among a handful of people in a small coffee shop in Hattiesburg, Mississippi. The company's leaders are equally proud of the fact that the original name for the company—Long Distance Discount Calling (LDDC)—came from the waitress. As with all self-perpetuated myths, whether true or not, this one serves a

---

[8] *Forbes*, 26 February 1996.   [9] *Fortune*, 26 February 1996.
[10] *Forbes*, 8 October 1997.   [11] *Qwest: Annual Report*, 1998, p. 32.
[12] *Qwest: Quarterly Report*, 30 June 1999, p. 14.

purpose, in this case to highlight the distance the company had travelled from small and humble beginnings. The contrast with the world's blue-blooded incumbents—such as AT&T, BT, Deutsche Telekom, FT, and NTT—was implicitly left for the observer to contemplate.

In 1999, only some sixteen years later, an extremely short period in the average life of the average large corporation, MCI-WorldCom had a market capitalization significantly higher than that of the company that traced its origins back to Alexander Graham Bell, namely AT&T. Leading financial analysts were proclaiming that with its unparalleled combination of assets, its global 'footprint', and the range of markets it was addressing it had become a 'must own' stock, in the same league, in terms of its size and forecast growth in earnings, as Microsoft, Coke, Disney, Wal-Mart, and Merck. Not only, they argued, was MCI-WorldCom the major global threat to the world's Big Five incumbents—AT&T, BT, Deutsche Telekom, FT and NTT—the company was better positioned from a competitive point of view than all of these rivals.

In 1985 Bernard Ebbers, an early investor in the company, became CEO. Born in 1941 in Edmonton, Alberta in Canada Ebbers dropped out of the University of Alberta and worked for a while as a high school basketball coach. Later he went to Mississippi College on a basketball scholarship. After college he became one of the owners of a small chain of motels.

To understand the growth of MCI-WorldCom it is necessary to examine its early business activities. As already mentioned, in September 1983 a group of people met in Hattiesburg, Mississippi to discuss the establishment of a start-up company to offer discounted long distance calling, initially in the state of Mississippi. Although there is some confusion regarding who attended the meeting,[13] more important than the precise attendees were the changing circumstances of the US Telecoms Industry that opened a window of opportunity for those entrepreneurs quick enough to spot it. More specifically, a month earlier, in August 1983, US District Judge Harold Greene had given final approval to AT&T's divestiture of its local operations, the Bell Operating Companies. This measure, long opposed by AT&T, opened competition in the US long-distance market for the first time. Two companies—MCI and Sprint—quickly became the major facilities-based competitors to AT&T in the US long-distance telephone market.

However, would-be new entrants into this market did not only have the option of building their own network facilities and competing with AT&T, as did

---

[13] On MCI-WorldCom's home page, in *Supplement to 1998 Annual Report*, it is stated that, 'In September 1983 businessmen Murray Waldron and William Rector met in a Hattiesburg coffee shop to discuss a business plan for a long distance reseller....And Bernie Ebbers, an early investor, became CEO in 1985'. However, in the company's earlier home page it is stated that in September 1983 'At a Hattiesburg, Miss., coffee shop, Bernard J. Ebbers, Bill Fields, David Singleton and Murray Waldron work out details for starting a long distance company', making no mention of William Rector. Furthermore, it is stated that 'Murray Waldron is named first president of LDDS. Bill Fields is named chairman of the board'. (http://206.65.84.57/timeline.html).

MCI and Sprint. An alternative mode of entry involved becoming a 'reseller', that is buying capacity from AT&T at a wholesale price and reselling this capacity to long distance customers at a price below that currently being charged by AT&T. Essentially, this operation involves arbitrage, buying a commodity in a cheaper market and selling it in a more expensive one. In this way the group that met in Hattiesburg, Mississippi in September 1983 began their company—LDDC[14]—without their own facilities, by using AT&T's network capacity and marketing their long-distance service under their own corporate brand at a price that was sufficiently lower than AT&T's to encourage long-distance callers to switch to LDDC. By July 1999, as shown in Table 6.2, WorldCom had become the fourteenth most valuable company in the world.

LESSONS REGARDING THE DYNAMICS OF THE TELECOMS INDUSTRY

In this section we return to the questions posed at the beginning of this chapter. What backgrounds did the new entrepreneurs come from? Two of the entrepreneurs—Gary Winnick of Global Crossing and George Soros of GTS—were essentially successful financial investors who saw the potential for profit in the new liberalizing Telecoms Industry.[15] Two of the entrepreneurs were essentially business people who had made their fortunes in other business areas and were then enticed into telecoms by the lure of profit. Philip Anschutz had made his fortune in oil and railroads before establishing Qwest and Walter Scott made his in construction and mining before setting up Level 3. In some senses, Bernard Ebbers is the odd man out. As a small investor in a small chain of motels he was an early investor in the company that was to become WorldCom, becoming its CEO in 1985. His background, accordingly, was altogether more humble than those of his counterparts in the other four companies.

How much prior knowledge did they have of the Telecoms Industry? The answer is not very much. Indeed, none of these new entrepreneurs had any significant experience in telecoms (with the partial exception of Winnick of Global Crossing who had played a role in the financing of MCI, the main long-distance challenger to AT&T, when he was with Drexel Burnham Lambert, although this left him with little knowledge about the industry).

This raises a puzzle: How could these entrepreneurs enter an industry about which they had so little knowledge? The answer is three-fold. First, they did not need much detailed knowledge of the Telecoms Industry in order to spot the opportunities for profit that emerged with the liberalization of telecoms. They knew that as a legacy of the monopoly period, telecoms services were

---

[14] The name of the company was later changed to Long Distance Discount Service, LDDS.
[15] The same is true, as we saw briefly, of Fidelity which founded Teleport in the US (later sold to AT&T) and COLT in the UK.

expensive. They also knew that new technology and efficient organization could give significant cost advantages to new entrants in comparison to the incumbent. Indeed, in the case of WorldCom that began as a reseller of AT&T's network capacity, new technology was not even necessary in the first stage of its growth.

The second part of the answer is that the network and business knowledge that was needed could be purchased on the labour market, especially by a profitable company with a rising share price. In this way Anschutz got his Joseph Nacchio (formerly head of consumer products at AT&T), Winnick got his Annunziata (also formerly with AT&T), and at lower levels all of the new companies got their skilled staff from other telecoms companies.

The third part of the answer is just as important. This is that they were able to buy much of the knowledge that they required to set up and run their telecoms networks in 'black-boxed' form from specialist technology suppliers, such as Nortel, Lucent, and Cisco. Indeed the new entrepreneurs did not even feel they had to do much of their own R&D. Their technology needs were simply outsourced from specialist suppliers who on occasion even provided some of the human skills that were needed.[16]

From where did they get the initial capital that was needed? Here too the answer is different for the different companies. For Qwest, Level 3, and GTS the initial capital had already been accumulated through the prior business activities of the entrepreneurs, Anschutz (in oil and railroads), Scott (in construction and mining), and Soros (financial investment). For WorldCom and Global Crossing, however, the initial capital had to be earned from scratch. Ebbers and colleagues were able to rely on being early-movers into the arbitrage opportunity provided by US telecoms liberalization, while Winnick could take advantage of the scarcity of capacity that existed at the time on trans-Atlantic cables and the new technology that increased capacity on new cables. Once their initial profits 'kicked in', all the new entrepreneurs received a substantial boost from the appreciating value of their shares provided by the stock market during the Telecoms Boom. In this way they were able to establish a virtuous circle whereby increased earnings gave them a more highly-valued acquisition currency, which enabled them to make more acquisitions, which further improved their earnings, etc. Until late 2000, that is, when the fundamental forces driving the Telecoms Industry began to catch up with them.

What does the experience of these new entrepreneurs and their new entrant companies teach us about the dynamics of the Telecoms Industry in the Internet Age? There are several salient points. First, it tells us about the significant opportunities for profit that were opened up by liberalization and technical change. Second, it tells us how low the barriers to entry were in an industry where all the important inputs could be purchased on well-operating

---

[16] The question of the role of R&D in the incumbents and new entrants is examined in detail in Chapter 8.

Table **6.4**. Market capitalization of selected US telecoms operators

Company	12 July 1999 ($ bn)	29 November 2000 ($ bn)	29 January 2002 ($ bn)
AT&T	186	70	65.17
Global Crossing	20	13	0.12
Level 3	27	12	1.79
Qwest	30	68	20.56
WorldCom	152	44	35.51

markets, including the capital markets that provided the means to purchase these inputs. Third, it tells us how important the activities of the new entrepreneurs were in creating an aggressively competitive environment, particularly in long-distance and international services. Not all of these entrepreneurs were already rich. Some, like Ebbers and his colleagues, were able to become rich by spotting early the opportunities that had become available.

STOCK MARKET PERFORMANCE—2002

By the end of January 2002, however, the fortunes of these new entrants had deteriorated significantly. This is evident from Table 6.4.

As Table 6.4 shows, between November 2000 and January 2002 the fortunes of Global Crossing, Level 3, and Qwest had declined significantly (though over this period AT&T and WorldCom were less badly affected). For GTS the situation was far worse; the company declared bankruptcy in June 2001.

Figure 6.1 compares the change in the share price of the incumbent, AT&T, and the two major new entrants, WorldCom and Qwest for the period 1997–2002.[17]

Figure 6.2 compares the share prices over the same five-year period for Level 3, Qwest, and WorldCom.

Finally, the proportionate change in share price over the five-year period is shown in Table 6.5 for the four companies.

As Table 6.5 makes clear, Qwest was the only one of the four companies to record a positive share price increase over the five-year period, 1997–2002. Qwest's performance is to a significant extent due to its acquisition of US West, a former Baby Bell with an important local access network. By contrast, the other three companies were more dependent on long-distance and international revenues.

Like GTS, however, Global Crossing, the remaining one of the five new entrants examined in this chapter, was hit by disaster. On 28 January 2002, Global Crossing filed for protection under Chapter 11 of the US bankruptcy

---

[17] This updates the similar graph in Chapter 1.

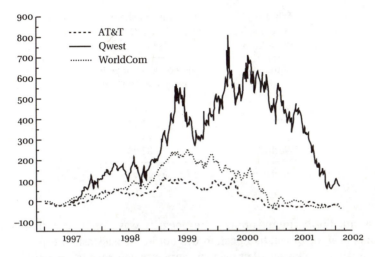

Fig. 6.1. Five-year share price comparison: AT&T, Qwest, WorldCom.

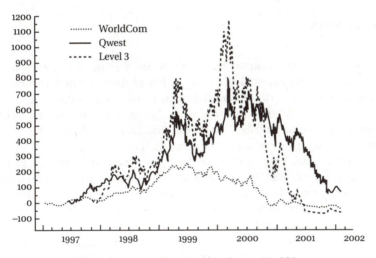

Fig. 6.2. Five-year share price comparison: Level 3, Qwest, WorldCom.

code.[18] The anticipation, according to the *Financial Times*, was that 'existing equity holders and preferred stockholders would be wiped out [by the bankruptcy]—as presumably would chairman Gary Winnick's remaining stake'.[19] Equally astounding on 30 April 2002 after a profits warning resulting from reduced internet earnings shocked Wall Street, Ebbers stepped down as CEO of

---

[18] The bankruptcy filing, however, did not include Asia Global Crossing, a separately quoted subsidiary that included Microsoft and Softbank as major shareholders.

[19] *Financial Times*, 28 January 2002.

Table 6.5. Percentage change in share price, 1997–2002: AT&T, Level 3, Qwest, WorldCom

Company	Percentage change in share price, 1997–2002
AT&T	−12.23
Level 3	−54.33
Qwest	+76.43
WorldCom	−32.08

Source: Bloomberg, 29 January 2002.

WorldCom. Furthermore, on 17 June 2002, it was announced that Joseph Nacchio, head of Qwest, had been forced out in the backlash sweeping the US in the wake of the Enron affair over lax corporate governance standards.

CONCLUSION

The main aim of this chapter has been to analyse the role played by the entrepreneurs who pioneered the entry of five of the main new entrant telecoms operators in the US. It was shown that not only did this investigation throw light on the different modes of entry that the entrepreneurs followed in entering the Telecoms Industry. The investigation also revealed much about the dynamics driving this industry in the Internet Age.

It was seen in this chapter that the fortunes of these five new entrant companies were very different, both relative to the incumbent, AT&T, and relative to one another. Ultimately, however, all five were, ironically, victims of their own collective success as was the incumbent (as shown in Chapter 3). The very success of their entry added significantly to capacity and increased the intensity of competition that overtime negatively affected both revenues and profits. The U-turn in investor sentiment as Telecoms Boom turned to Bust added greatly to these negative effects, deepening the downturn in their fortunes. Only time will tell who the longer-term survivors will be and how well they will do.

7

COLT: An American Telecoms Operator in Europe

Until 2000 City of London Telecommunications Ltd (COLT), established by Fidelity the largest US mutual fund, was widely regarded as the most successful new entrant network operator in Europe and as one of the main threats to the British incumbent, BT. By 2002, however, although COLT remained a major player in Europe, the company's financial fortunes had declined significantly, illustrating the analyses of the Telecoms Boom and Bust provided in Chapter 1 and the dynamics underlying the evolution of the Telecoms Industry examined in Chapter 2.

Why was COLT such an outstanding performer and, equally important, why did its fortunes deteriorate so rapidly? What does the future hold for COLT? These are the main questions examined in this chapter.

INTRODUCTION

City of London Telecommunications Ltd was established in 1992 by Fidelity Capital, a subsidiary of the largest US mutual fund. The company started with only seven employees. In April 1993, the company received its UK national public telephone operator licence and in October 1993 it connected its first customer. In December 1996 COLT had its IPO. Its shares were floated at 69 pence and on the first day started trading at around 76 pence. At the end of 1996 COLT's market capitalization was £256 million. In 1998 COLT joined the FTSE 100 index and in this year it was the London Stock Exchange's best

This chapter is based on interviews with key COLT staff, including Paul Chisholm, President and CEO of COLT until January 2001. However, the analysis contained in this chapter should be attributed neither to Mr Chisholm nor to COLT staff. I would also like to thank Ian Duff for able research assistance.

performing share. At the end of 1999 COLT's market capitalization was £21.2 billion. In the first quarter of 2000 COLT's share price reached an all-time high of £40.78. After the company released its third quarter results in 2000 the *Financial Times* expressed the view that 'There is little to diminish the compelling argument that COLT is the best-performing alternative telecoms operator in the European market'.[1]

However, when COLT issued its second quarter results in August 2001 it became clear that the company was beginning to be negatively affected by the slowdown in European economic growth, by a weakening in the wholesale market, and by the falloff in business occasioned by the collapse of many dotcom companies. In August 2001 Moody, the credit rating agency, cut its evaluation of COLT from 'positive' to 'stable'. The rapidity with which COLT's financial fortunes declined thereafter was astounding. In September 2001 COLT was removed from the FTSE 100 index (along with telecoms companies Energis, Marconi, and Telewest). On 5 October 2001 COLT's market capitalization had fallen to around £632 million. On the same day the company's share price traded at around 90 pence, a fall of about 98 per cent from the share's high (even though the 90 pence reflected a significant appreciation after Fidelity announced that it would underwrite a significant capital injection into the company).

The aim in this chapter is to explain this significant U-turn in COLT's financial fortunes, drawing on the insights provided in Chapters 1 and 2.

COLT AND THE CONSENSUAL VISION

COLT fitted perfectly into the Consensual Vision of the Telecoms Industry analysed in Chapter 1. Indeed, COLT provided one way in which the Consensual Vision was diffused as a cognitive framework from the US to the UK and Europe. This is apparent from the following quotations that come from a report on COLT by the financial institution Salomon Smith Barney carrying the names of Andrew Harrington and Richard Sloane and published in January 1998. Harrington and Sloane were London-based members of Salomon Smith Barney's Global Telecom Team headed by Jack B. Grubman in the US. The first paragraph of the report states 'COLT is a competitive local exchange carrier (CLEC) entering Europe's rapidly liberalising telecoms markets. Experience in the US suggests that such CLECs can gain significant market share'.[2]

In the second paragraph of the report the authors explain why they see COLT as a 'premier new entrant opportunity': 'The total European telecoms market is at least as big as the US, but is still dominated by the incumbent

---

[1] *Financial Times*, 8 November 2000.
[2] Andrew Harrington and Richard Sloane, 'COLT: A Premier New Entrant Opportunity', *Salomon Smith Barney*, Global Equity Research, Telecommunications, London, 22 January 1998, p. 3.

former-monopoly carriers. All incumbents inevitably have an established operating structure and technology, which hinders their flexibility in product offerings and customer service. New entrants, including COLT, have significant competitive advantages because they have a flexible service and marketing structure using a broadband network platform'.[3]

On the second page of the report the authors point out that COLT, like other new entrants, has a crucial advantage *vis-à-vis* its rival incumbent, BT, in addition to its more competitive 'operating structure and technology' referred to in the last quotation: 'In building and operating its networks, COLT has focused on large European financial centres. These centres usually contain a disproportionate amount of the national telecommunications spend over a relatively small portion of the incumbent's network. For example, perhaps 40% of UK and French telecommunications revenues are in London and Paris, which means COLT can address these revenues with a fraction of the total assets that BT and France Telecom have'.[4]

Very similar arguments were put forward by many of the other leading financial institutions that produced reports on COLT at this time, justifying the reference in Chapter 1 to a Consensual Vision regarding the relative competitiveness of the new entrant telecom operators compared to the incumbents.

COLT'S ENTRY STRATEGY

Why was COLT able to enter the UK (and later the European) telecoms market so easily and successfully? The answer to this question reveals a great deal, not only about COLT as a company, but equally importantly about the changing structure and dynamics of the Telecoms Industry in the Internet Age as explained in Chapters 1 and 2.

There were five planks in COLT's entry strategy.[5] These are shown in Table 7.1.

The first plank was, and remains, to focus on Europe. The reason was that in the eyes of COLT and its major investor, Fidelity, Europe offered particularly attractive investment opportunities since, relative to the US, it was a latecomer to the telecoms liberalization process. Indeed, it was only on 1 January 1998 that the European Union formally liberalized all its telecoms markets. This meant that the incumbent telecoms operators' prices were still high relative to the US, giving COLT additional opportunities to exploit.

Early market entry constituted the second plank. As an early bird, COLT would be able to catch several worms that might not be available to later

---

[3] Harrington and Sloane (1998) op.cit., p. 3. It is worth stressing that this quotation generalizes about 'all incumbents' and all new entrants, of which COLT is one example.

[4] Harrington and Sloane (1998) op.cit., p. 4. Once again, the statement is a generalized statement, implicitly applying not only to COLT, but also to other new entrants.

[5] For further details on COLT's strategy, see *COLT Telecom Group plc, Stock Offering of 7,200,000 Ordinary Shares*, 24 July 1998, pp. 6–7, henceforth referred to as *COLT Stock Offering*.

Table 7.1. COLT'S entry strategy

Focus on Europe
Enter markets early
Operate through local companies
Provide a wide range of high quality telecoms
    products and superior customer service
Focus on large end users and carrier customers

comers. As an early entrant COLT would have few new entrant competitors, leaving a relatively large price differential between it and the incumbent. This price advantage could be used to drive the effort to secure initial large and well-known customers who, in turn, could be used to advertise the merits of switching to the new kid on the block. The creation of a focused, fast, flat, and flexible organization would allow COLT to understand the telecoms needs of business customers, and respond to them, more effectively than the incumbent.

Furthermore, as an early new entrant COLT would face lower entry barriers than would later comers. Quite simply, there is a limit to the number of local network providers that will be given permission to dig up streets and pavements and install their equipment in buildings. In addition, according to the company, '[COLT] believes that later entrants in markets where [COLT] is an early entrant will face higher barriers to entry in the short term, including time-consuming negotiations for local connectivity and the technically difficult, capital intensive build-out of a network in environments of multiple local standards'.[6]

The third element in COLT's strategy involved establishing in each of the European countries in which it operates decentralized local companies staffed by local people. This form of organization was intended to simplify the overall structure of the company, making it less hierarchical and more responsive to local needs, including the needs of local employees. It recognized that local-ization may facilitate better access to local authorities for the permission that is essential to build the company's local access networks and would allow for a better understanding of local customer needs.

According to the fourth plank, relatively higher quality and superior customer service would also help to give COLT a competitive edge in its battle with the incumbent. Here its latest state-of-the-art technology would assist it against the incumbent impeded by its old-technology legacy network. For example, COLT's SDH/SONET network, the first all-SDH/SONET network in the UK, facilitated a lower probability of faults occurring, and, if they were to occur, a shorter time for repairing them, as a result of the self-healing properties of these networks.[7]

---

[6] *COLT Stock Offering*, p. 6.
[7] SDH, synchronous digital hierarchy (SONET in the US), is a set of standards that facilitate optical communications transmission. It allows for the interoperability of incompatible vendor's equipment and significantly improves both the reliability and the flexibility of

Finally, COLT would begin by focusing on large customers as well as other network operators buying in the wholesale market. As already noted, large customers, such as the financial institutions and media companies that COLT at first targeted, are frequently concentrated in fairly small parts of a country's major cities. This means that they can be reached by a relatively small network, implying a lower cost of connection per customer than that faced by the incumbent who, through universal service regulation, is obligated to cover not only the whole city but also the whole country. With its local access networks installed in the major European cities, COLT would also be in a strong competitive position to sell to other network operators.

COLT'S ENTRY PROCESS

COLT was well-placed to implement its entry strategy. One of the reasons was that it was able to learn from the knowledge that had been accumulated by Fidelity in the US.

Learning from the US Experience: Fidelity, Teleport, and COLT

It is significant that COLT was established—not by a telecoms operator, a utility, or an industrial group as were most of the other new entrants in Europe—but by a financial institution, Fidelity Capital, a subsidiary of the US's largest mutual fund. Financial institutions, in exercising their capital market functions, normally provide funds to borrowers without securing controlling ownership rights over those borrowers. However, Fidelity Capital was established with the specific purpose of investing Fidelity's earnings, taking advantage of its parent's superior market information, capital availability, and reputation in financial markets. In investing these earnings Fidelity was quick to spot the new profitable opportunities that emerged when telecoms markets in the US began to liberalize from 1984 with the divestiture of AT&T and the beginnings of competition in telecoms services.

COLT was not Fidelity's first investment in the Telecoms Industry. In 1984 Teleport was established by Fidelity with its initial operation in New York and with Robert Annunziata, a former AT&T employee, as its head. One year later the company began providing private line and other dedicated access services in New York. In 1987 Teleport attempted to generalize its business model by opening facilities in Boston. In 1988 Teleport was recognized as a CLEC by the

---

telecoms services. In its sales and publicity material COLT went to great lengths to stress the competitive advantage that it enjoyed in the local services market—particularly relative to the incumbent, BT—from its all-SDH network.

Massachusetts authorities. In the same year the Teleport Communications Group (TCG) was established as an umbrella organization to take charge of Fidelity's operations in various US cities.

In July 1994 the New York Public Service Commission authorized TCG to become the US's first CLEC providing local calling services to businesses within the New York metropolitan area through six digital switches. By the beginning of 1998, TCG had more than 3,000 employees and was one of the US's major suppliers of competitive local communications services (in competition with the Baby Bells, the Bell regional holding companies that separated from AT&T at the time of divestiture and retained *de facto* monopoly positions in local services). TCG's fibre optic network covered more than 250 communities in the US from coast to coast, including 66 of the country's major cities. Together with Merrill Lynch, Fidelity Capital was one of the founding partners in TCG.

On 8 January 1998 AT&T announced a merger agreement with TCG for an all stock transaction valued at $11.3 billion. In justifying the agreement, AT&T stated that 'Joining forces with TCG will speed AT&T's entry into the local business market'. At the time of the merger three of the largest US cable companies— Cox Communications, Comcast Corporation, and TeleCommunications Inc. (TCI)—having acquired TCG, held 95 per cent of the voting power in TCG and 66 per cent of its equity, with the remaining 34 per cent publicly traded on the NASDAQ market. Later on in that year AT&T went on to further extend its activities in local telecoms through its acquisition of the cable TV company, TCI.

COLT was established by Fidelity Capital in 1992 with Paul Chisholm as Managing Director, leading an initial team of only seven employees. In 1995 Chisholm became President of COLT and in 1996 also the company's Chief Executive Officer. Chisholm brought the experience and learning of Teleport with him to London. With earlier employment in the New England Telephone & Telegraph company, in AT&T, and as Vice-President-Telecommunications in the Shawmut Bank in Boston, Paul Chisholm served as Vice President and General Manager of Teleport Communications Boston from 1988—the year in which Teleport began operations in Boston after its success in New York—to 1992.

The latter four of COLT's five planks in its entry strategy (see Table 7.1) were derived directly from Fidelity's (and Chisholm's) experience with Teleport in the US. This provided an important transmission process, based on learning, for the new knowledge that was being accumulated in the US as a result of telecoms liberalization to be transferred to Europe.

Climbing over the Barriers to Entry

With the accumulated knowledge of the Teleport experience to draw on, COLT was well-positioned to deal with the most important barriers to entry. These included money, management and skilled labour, and technology.

Money. The supply of capital was obviously less of an entry barrier for COLT than it was for many other new entrant telecoms operators. As a financial market operator, and one that had already accumulated first-hand experience in the Telecoms Industry, Fidelity was well-placed to take care of COLT's financial needs (a role that it continued to play through 2001 when it bailed the financially beleaguered company out with a further crucial injection of capital). As a result COLT was able relatively easily to surmount the capital entry barrier.

Management and Skilled Labour. When it came to getting the management and skilled labour that COLT needed in order to become a new entrant network operator the company was able to turn to the labour market and through it make use of some of the knowledge that had been accumulated by other telecoms operators, including incumbents. This helped COLT to deal with the management and skilled-labour barrier to entry.

Mention has already been made of Paul Chisholm's earlier experience with New England Telephone & Telegraph and AT&T in addition to Teleport. Ronald Dillon, Director of International Operations for COLT, before joining the company worked with BT in London for 22 years in various customer services and operational management positions. E. Jonathan Watts, Managing Director of COLT Telecommunications, from 1990 to 1994 had experience with another incumbent, namely the US Baby Bell, BellSouth, for whom he worked as Managing Director for European Operations. Horst Enzelmuller, Managing Director of COLT's German operations, formerly held senior management positions with BT in Germany and France and in Germany was responsible for negotiating BT's joint venture with Viag. Similarly, Claude Olier, Managing Director of COLT's French subsidiary, was formerly Chief Executive of Eunetcom, a joint venture established by France Telecom and Deutsche Telekom.

In dealing with the management and skilled-labour entry barrier, COLT was also assisted by its 'F4' form of organization, that is the fact that it was focused, fast, flat, and flexible. The organizational characteristics and capabilities were facilitated by the small size of the company, its ability to address such a large proportion of the UK business market with a relatively small number of employees. This is shown in Table 7.2.

In March 1999 BT employed around 125,000 people worldwide and in 2000 the figure was 137,000.[8] In 1998 it was estimated that by 2003 COLT would have a market share of around 5 per cent of the UK market for switched business calls.[9]

Technology. In getting access to the technology necessary to enter the Telecoms Industry and compete with both incumbents and new entrants COLT, like all network operators, has been able to make use of the vertical specialization in

---

[8] BT, Annual Reports, 1999 and 2000. [9] Harrington and Sloane (1998) op.cit., p. 9.

Table 7.2. COLT's employment

Category	1998	1999	2000
Engineering and operations	578	1,083	1,840
Sales and marketing	258	414	790
Administration	236	424	650
Total	1,072	1,921	3,280

Source: COLT Annual Report, 2000, p. 51.

this industry and buy the equipment and services of specialized telecoms equipment suppliers.[10] In this way COLT has been able to use the technology market to overcome the technology barrier to entry.

In developing its first optical fibre network in the City of London COLT entered into contractual agreements with several major specialist technology suppliers. A particularly important agreement was concluded with the Canadian telecoms equipment supplier, Nortel. It was Nortel that supplied COLT with the switches—particularly the Nortel DMS 100 switches—and some of the transmissions systems that it needed. The British equipment supplier, GPT (later Marconi), long the major supplier of telecoms switches to BT (British Telecom), supplied COLT with the SDH transmissions sytem that became one of the first fully SDH networks in the UK. Nortel, GPT, and Nokia supplied COLT with electronic devices while Pirelli supplied the optical fibre cable. When COLT later started building its network in Germany it turned to Siemens as a second major supplier of switches, taking advantage of Siemens' strong position in Germany.

In addition to its physical telecoms network, COLT also depends on billing and customer service information systems that the company sees as a 'core capability necessary to record and process the data generated by a telecommunications service provider'.[11] These systems assist COLT in areas such as project and fault management, the integration of new customers, improving the least-cost routing of traffic on the company's network, and generally improving its quality of service to customers. In the area of billing systems, for example, COLT uses the systems of suppliers like Ericsson, Hewlett Packard, and Texas Instruments.

COLT, like the other new entrant network operators, differs strongly from the incumbents in that it has made the decision not to develop significant in-house R&D capabilities but in effect to outsource its R&D together with its telecoms equipment requirements. In order to govern the process of R&D and equipment procurement, COLT has developed long-term obligational relationships

---

[10] The importance of vertical specialization in the Telecoms Industry and the role of specialist telecoms equipment suppliers has been analysed in Chapters 1 and 2.

[11] *COLT Stock Offering*, p. 47.

with a few selected key suppliers. Being a relatively large customer with a rapidly developing pan-European network COLT has managed to use its financial muscle to ensure that its R&D and equipment requirements are satisfactorily met. The question, however, of whether the absence of a significant internal R&D capability constitutes a longer term disadvantage for COLT in its competitive rivalry with the R&D-performing incumbents, who spent significant annual amounts on R&D, remains a controversial issue.[12]

COLT'S EVOLVING CORPORATE STRATEGY

As was made clear earlier, COLT's initial strategy involved starting as a local fibre optic telecom company. In other words, the company began by building its own local area optical fibre networks in core business districts in London and then other major European cities. These networks were then linked in order to develop a substantial pan-European network. Company strategy, however, determined that no attempt would be made to extend the network outside Europe.

As already noted, a strength of this strategy was that it took advantage of the inherent bottleneck nature of the local access market. While it is relatively easy to secure rights of way and develop long-distance networks (e.g. by building them alongside railways—as Qwest has done in the US—or roads), it is far more complex and costly to develop local access networks, particularly in crowded key business districts in the main urban areas. It is precisely for this reason that the local access market is where the incumbents have tended to retain their monopolistic control whereas in the long-distance and international markets they have been forced to compete more fiercely.[13] Although over time competition may well come increasingly from operators using alternative technologies—such as cable TV, mobile phones, fixed radio access, satellites, and lasers—COLT's profit margins may well, even in the longer term, benefit from the 'bottleneck proneness' of local access. In short, COLT may be able to continue benefiting from a scarcity rent.[14]

However, under Chisholm COLT has also attempted to move up the value chain to become, in his words, 'an integrated local, long distance and hosting services company'. By 'long-distance' he is referring primarily to the connections

---

[12] See Chapter 8 for a more detailed analysis of the changing role of R&D in the Telecoms Industry.

[13] A good deal of evidence to support this statement is to be found in the chapters on AT&T, BT, Deutsche Telekom, France Telecom, and NTT. Similarly, it is no accident that the former Baby Bells in the US, particularly Verizon and Cingular, have benefited relative to companies such as AT&T and WorldCom as a result of their strong base in the local access market.

[14] It is no accident, as can be seen from the figures on the incumbent network operators given in Chapters 3–5, that it is precisely in area of long-distance and international voice and data services, where there are least bottlenecks, that prices and profitability have fallen most rapidly.

between COLT's metropolitan area networks in European cities. By hosting services he is referring to the web-hosting data centres, which COLT refers to as Internet Solution Centres, that the company has established in several cities.

ACHIEVEMENTS TO MID-2001

COLT's IPO was held in December 1996. Its shares, floated at 69 pence, started trading at around 76 pence. By the end of 1996 COLT's market capitalization was £256 million. In 1998 COLT joined the FTSE 100 index and in this year it was the London Stock Exchange's best performing share. At the end of 1999 COLT's market capitalization was £21.2 billion. In the first quarter of 2000 COLT's share price reached an all-time high of £40.78.[15]

In May 2000 Forbes magazine concluded that, 'COLT is the crown jewel of Fidelity's 50-plus venture investments'.[16] Also an achievement of sorts, COLT could boast (if it chose to) that its President and CEO, Paul Chisholm was the highest-paid executive in the UK in 1999. In this year the value of his remuneration package, including both salary and share options, amounted to £146 million. According to the calculations of Fortune magazine, Chisholm in 1999 was the 278th richest person in the UK, a reasonable achievement for someone who arrived in the country to start COLT, together with six employees, only in 1992. In August 2000, however, Chisholm announced that he would step down at President and CEO, citing as one of the main reasons that he was tired of commuting twice a month to Boston to see his family.[17] On 11 January 2001 he stood down, retaining a position on the Board of Directors.

By 2001 COLT had completed its 15,000 kilometre inter-city network which included operational networks in twenty-seven cities—by and large the commercially most important cities in Europe—in eleven countries. Although, as mentioned and as still embodied in the company's name, COLT began its operations in London, by the end of 2000 Germany accounted for slightly more of its total turnover, with the UK following a little way behind. France was in third position, accounting for about one-third of the total turnover of Germany.

---

[15] At many points in its history, COLT's share price benefited from rumours that the company would be acquired by a deep-pocketed predator. Indeed, Salomon Smith Barney, in a report on the company dated January 1998, argued that one of the factors benefiting COLT's valuation was that it was a 'likely take-out candidate as the global consolidation continues'. Nevertheless, the analysts warned, 'COLT does have one big weakness: its lack of global assets. This will increasingly become a disadvantage in an era where the ability to offer a seamless global product will be crucial in gaining significant telecom revenues from multinational corporates. However . . . Fidelity Capital, COLT's main shareholder, is a financial shareholder and we believe could therefore be willing to sell COLT at the right price' (Harrington and Sloane 1998, op.cit., p. 3). By the beginning of 2002 it appeared that none of these predictions were correct.   [16] *Forbes*, 15 March 2000.

[17] *Financial Times*, 9 August 2000.

Table 7.3. COLT'S profit and loss, 1998–2000[18]

	1998 (£'000)	1999 (£'000)	2000 (£'000)
Turnover	215,052	401,552	686,977
Cost of sales (including interconnect/network and network depreciation)	(172,351)	(323,605)	(566,904)
Gross profit	42,701	77,947	120,073
Operating loss	(34,130)	(65,321)	(84,913)
Loss (after taxation)	(55,602)	(101,261)	(116,860)

Source: COLT, Annual Report 2000, p. 42.

Currently, most financial analysts agree that COLT is probably the best of all the alternative network operators competing with the European incumbents.

CRACKS BEGIN TO APPEAR ...

As high as COLT was flying by early 2001 there was a structural weakness in its engines. In the heady world of the Telecoms Boom (analysed in Chapter 1) this weakness was barely audible above the roar of the turbos. However, as boom began to crumble into bust and the roar turned to a murmur so the weakness was more easily heard. This weakness is shown in Table 7.3.

As Table 7.3 makes clear, COLT's heady share price during the Telecoms Boom was based more on promise than performance. The reason was simple and indeed is the general rule for entrants into high fixed-cost industries such as telecoms. The problem stems from the fact that the high fixed costs mean that significant expenditure must usually be made *before* the compensating revenue streams provide relief. The Telecoms Boom, based on the Consensual Vision analysed in Chapter 1, provided the ideal situation in which to make good the gap between expenditure and revenue. Essentially, it was the bullish expectations of the financial market investors that plugged the gap. The Consensual Vision provided the rationale for believing that future earnings would be sufficiently great to justify paying a high share price now (or lending to the company at the going interest rate) and covering the risks involved.

... AND THE SHARE PRICE COLLAPSES

The problem, however, is that this logic ceases once Boom turns to Bust and expectations regarding future earnings become significantly more pessimistic. Then there is a flight to profit *now*. And this is precisely the problem that COLT

[18] For year ended 31 December.

faced, even though the fundamentals of the company's business remained as sound as they always were and financial analysts continued to agree that COLT offered a sound business proposition and was still the strongest of all the European new entrant alternative network operators.

However, in the second half of 2001 COLT also had to cope with reduced revenue as a result of the deepening economic recession in Europe and the collapse of the dotcoms. In August 2001 when COLT issued its second quarter results it became clear that the company was beginning to be negatively affected by the slowdown, by the failure of many dotcoms, and by a weakening in the wholesale market (i.e. the market selling capacity to other telecoms operators). In August 2001 Moody, the credit rating agency, cut its evaluation of COLT from 'positive' to 'stable'. The rapidity with which COLT's financial fortunes declined thereafter was surprising. The company's share price plummeted. In September 2001 COLT was removed from the FTSE 100 index (along with telecoms companies Energis, Marconi, and Telewest). On 5 October 2001 COLT's market capitalization was around £632 million. On the same day the company's share price traded at around 90 pence, a fall of about 98 per cent from the share's high.

The 90 pence price, however, included an appreciation that was made the previous day when Fidelity announced that it would underwrite an equity issue to raise an additional £400 million for COLT. In making this refinancing decision, Fidelity was committing itself to spending around four times more in this one injection than it had during its whole association with COLT over the company's nine-year history.[19] This despite the fact that Fidelity had taken a paper loss of about £13.5 billion in COLT since the Telecoms Bubble burst in mid-2000.[20] If ever proof was needed of Fidelity's conviction that, despite the Bust, COLT remained an attractive financial investment relative to all the other options, this was it.

WHAT THE ANALYSTS MADE OF IT ALL

Usually, as Boom turns to Bust, the financial analysts, in public at least, make little of it all but continue with their new valuations, leaving the questions begging for answers. Sometimes, however, a few necks are exposed.

One such neck was that of Morgan Stanley Dean Witter.[21] In answer to the dreaded question, 'Where were we wrong?', the company's analysts provide three answers:

1. Market sentiment was worse than we expected; which led to financing issues for many carriers, higher required rates of return and a short-term outlook.

---

[19] *Financial Times*, 4 October 2001.   [20] *The Observer*, 2 September 2001.
[21] Kennedy, P. *et al.*, 'COLT Telecom Group', 18 September 2001.

2. The carrier wholesale market deterioration was faster and deeper than we anticipated.
3. Value-added services growth is slower than predicted.[22]

Nevertheless, the analysts were adamant that their analysis of the 'fundamental outlook' was correct. The reason was that the attractions of 'the alternative carrier model' remained unchanged. They gave three reasons to support their contention:

1. The pricing umbrella provided by the incumbent still exists.
2. The telecom market is huge and still growing strongly.
3. Enterprise demand will drive growth as telecom becomes more of a productivity tool.[23]

Accordingly, they concluded that 'COLT is a long-term play and one of the best-positioned alternative carriers in Europe'.[24] Generally, although most other financial analysts have not been as forthcoming regarding past failures and while there are some differences in detail, this sentiment was shared by most of the analysts reporting on COLT in the second-half of 2001.[25]

CONCLUSIONS

The story of COLT's financial collapse (and that of many other alternative telecoms operators) presents an important conundrum. If the 'fundamentals' of the company, that is the viability of its business model, remain constant—as most financial analysts seem to agree—why has its stock market valuation fallen so much?[26] This question needs to be answered.

In theory, the present price of a company's shares (and therefore its market capitalization) is determined by the present value of its expected future earnings. If this is true, then market capitalization can only change if expected future earnings change and/or if the discount rate changes. However, in COLT's case expected future earnings would have had to have changed significantly to account for the substantial reduction in the company's market capitalization. In which case the company's 'fundamentals' have not remained constant. Or the discount rate has changed significantly, in which case there has been a substantial change, not in the fundamentals of the company, but in the time-preference of money determined in the financial markets.

---

[22] Kennedy, P. *et al.* (2001), op.cit., p. 3.   [23] *Ibid.*   [24] *Ibid.*
[25] By the end of January 2002, however, 'sentiment' in financial markets regarding so-called alternative network operators like COLT became significantly more pessimistic. As a sign of the times, Global Crossing (analysed in more detail in the last chapter), one of the best-known alternative operators in the US, on 28 January 2002 filed for protection under Chapter 11 of the US bankruptcy code. The pessimism regarding alternative operators sweeping global financial markets affected all the new entrants.
[26] In January 2002 COLT's share price varied from about 70 pence to around £1.50.

In either case it may be logically deduced that a fundamental change has occurred in the company and/or in the financial markets. Of course, this does not necessarily make COLT's shares less attractive as a financial asset. If the same revaluation has occurred with all financial assets, COLT's *relative* attractiveness will remain the same. And it is relative attractiveness that will drive demand and supply of a company's shares.

Tempting though it may be, it would also be wrong to attempt to separate COLT's real business activities and prospects from their financial values. The reason is that the change in what the financial analysts refer to as 'market sentiment' has had an important real impact on COLT's business. For example, by increasing the real interest rate that must be paid in order to borrow to fill the gap between its expenditure and revenue, financial markets have forced COLT to reduce its rate of investment. Financial markets have thus increased the amount of time that will be needed to address the markets in Europe that COLT has selected. In short, the financial bust has had a significant impact on COLT's business and in this sense, therefore, it cannot be said that the company's 'fundamentals' have remained unchanged.

Turning more specifically to the details of COLT's business, there are several key determinants of the company's future fortunes that must be highlighted. First is the growth of overall demand for the company's products and services which is likely to be robust as a result of the growing usage of data, information, and the Internet in Europe. Second, as stressed in several places in this chapter, COLT is well-placed in the 'bottleneck-prone' local access market. Although in the longer term threats may very well come from technological alternatives to optical fibre, COLT is likely to continue to benefit from the barriers faced by would-be competitors in this local market. Third, it may well be that the bust, by speeding the processes of exit and consolidation, has helped COLT by reducing the number of competitors in its markets. To the extent that oligopolistic competition represents the long-run outcome for the Telecoms Industry, COLT is likely to be a beneficiary.

The fourth point, however, is that the second and third elements in Paul Chisholm's description of COLT as an 'integrated local, long distance and hosting services company' may well come under greater competitive threat than the first element. Simply put, there are unlikely to be the same entry barriers in the long-distance and hosting markets that there are in the local access market. To the extent that there is greater competition in the latter two areas, COLT's 'scarcity rent earnings' will be lower in these areas.

Combining these points with the changed circumstances in financial markets already referred to, it is likely that COLT will be in a reasonable position in the longer term, provided it can weather the rest of the short term financial storm. Certainly, Fidelity believes it has a good chance and as a major financial market player, and now COLT's major shareholder, Fidelity has some influence over the outcome.

8

The Changing Knowledge Base: Co-evolving R&D in Telecoms Network Operators

Some of the most important innovations that have driven, and still drive, the global Infocommunications Industry have originated in the R&D laboratories of the incumbent network operators. These include famous laboratories such as AT&T's Bell Laboratories, NTT's Electrical Communications Laboratories, BT's Martlesham Laboratories, and France Telecom's (FT's) CNET Laboratories. Over time, however, as a result of the process of vertical specialization analysed in Chapters 1 and 2 of this book, the role of R&D in these laboratories and companies began to change dramatically. Specialist telecoms equipment suppliers—such as Nortel, Lucent, Cisco, NEC, Fujitsu, Ericsson, and Nokia—began in some areas to take over much of the R&D formerly done in the incumbent network operators' laboratories. Furthermore, the latter changed fundamentally not only the proportion of their resources allocated to R&D but also their whole philosophy regarding the importance of R&D and its organization. The result has been a dramatic shift in the knowledge base of the entire Telecoms Industry, with important implications for the dynamics of the industry. These significant changes are analysed in this chapter.

INTRODUCTION

The R&D Conundrum

A key element in understanding the co-evolution of companies, industry, and R&D is what I will call the 'R&D Conundrum'. This conundrum may be expressed in the form of a question: Why allocate scarce company resources to R&D, whose benefits are uncertain and only accrue in the future? It is this conundrum that every chief technical officer (CTO) and company laboratory manager has to deal with, practically on a daily basis and often in the face of strong scepticism from CEOs and the heads of other corporate functions such as finance and marketing. It is the negotiation and resolution of this conundrum, under changing circumstance as the company and industry evolve, that determines the evolution of corporate R&D.

PERIODIZING THE EVOLUTION OF THE TELECOMS INDUSTRY AND R&D

How have the Telecoms Industry and the R&D function in this industry co-evolved? Tables 8.1 and 8.2 provide a summary of what is in reality a highly complex co-evolutionary process.

As shown in Table 8.1, the evolution of the Telecoms Industry may be divided (from the perspective of the beginning of the twenty-first century) into four periods. Period 1 is up to the mid-1980s when liberalization was first introduced in Japan, the UK, and US and their incumbents—NTT, BT, and AT&T—first began to face serious competition. Period 2 lasts from the mid-1980s to the mid-1990s. During this period new entrants begin to enter the Telecoms Industry, companies such as DDI, Japan Telecom, and Teleway Japan in Japan; Mercury and later COLT and Energis in the UK; and MCI, Sprint, and WorldCom in the US. During this period, although data communications makes its appearance and the Internet begins to be used by a growing number of people, it is voice communications that remains the most important activity, and source of revenue and profit, in the Telecoms Industry.

It is in Period 3, however, that several important radical changes begin to be felt. The Internet revolution begins to truly take off based on its triad of technologies: packet switching, Internet Protocol, and the World Wide Web. In 1995 Bill Gates acknowledges that henceforth it is the Internet that will dominate everything in the Infocommunications Industry. Corporate data communications become far more important. There is also a mass adoption of mobile communications that becomes one of the fastest growing segments in the industry. The boom in Internet, data, and mobile generates excess demand for capacity. In turn, capital markets, sensing the opportunity for above-average profits from the 'explosion in demand' for telecoms capacity, and in the context of a generalized bull stock market in the US and Europe, pour billions of dollars into the Telecoms Industry.

In Period 4, from 2000, the picture again changes dramatically. A combination of competition, massive investment in capacity, and technical change hits traditional sources of revenue. Incumbent telecoms operators acknowledge that their revenue and profits growth has been significantly hit by poor performance in fixed voice services (see Chapters 3–5). Surprising some, WorldCom, the

Table 8.1. Periodizing the evolution of R&D in the telecoms network operators

Period	Dates	Description	R&D Incumbents	R&D New entrants
Period 1	To mid-1980s	Pre-liberalization	Central research labs (CRL), arm's-length from businesses	None
Period 2	Mid-1980s to mid-1990s	Liberalization—new entry—competition	Linking CRL more closely to businesses	R&D outsourced
		Voice dominant		
Period 3	Mid-1990s to 2000	Internet boom from about 1995	Development increasingly moves into businesses	R&D outsourced
		Mobile boom from about 1997	Research remains in diminished CRLs	
		Excess demand for capacity	Reduction in long term and basic research	
		Massive support from capital markets, driven by bull market in US and Europe		
Period 4	2000 to ?	Competition + massive investment in capacity + technical change hit traditional revenues	CRL and research threatened as company's businesses are empowered and some companies break up	Some new entrants begin business-based R&D No CRLs Others continue to rely on outsourcing
		For many telcos new growth areas (Internet, data, mobile) do not compensate	Development moves entirely into businesses	
		For many telcos falling revenue and earnings growth		
		Bull turns into a bear		

Source: M. Fransman.

most successful of all the new entrants, announces in 2000 that it too is suffering from precisely the same problems. Bernard Ebbers, CEO of WorldCom, admits that he has failed both his shareholders, used to rapidly rising shareholder value, as well as himself.

This downturn in corporate fortunes, however, does not hit all telecoms companies. Some are fortunate to be more narrowly focused on the new growth sectors, and are not dependent on the slowing growth in fixed voice. The best

Table 8.2. Telecoms R&D, 1999

	1999 R&D spend ($'000)	Sales ($m)	R&D % sales	R&D per employee ($'000)
Telecoms operators				
AT&T	550,000	62,391	0.9	2.3
BT	556,037	30,163	1.8	2.6
Deutsche Telekom	701,611	35,552	2.0	2.2
FT	594,572	27,297	2.2	2.1
NTT	3,729,910	95,061	3.9	10.3
Specialist telecoms suppliers				
Cisco	1,594,000	12,154	13.1	47.1
Ericsson	3,877,196	25,214	15.4	22.9
Fujitsu	3,859,723	51,224	7.5	12.7
Lucent	4,510,000	38,303	11.8	18.3
NEC	3,382,483	46,495	7.3	13.3
Nokia	2,030,662	19,817	10.2	22.8
Nortel	2,908,000	22,217	13.1	23.5
Sectors				
Telecoms			2.6	
Automobiles			4.2	
Beverages			2.2	
IT hardware			7.9	
Media and photography			4.2	
Personal care			3.3	
Pharmaceuticals			12.8	
Software & IT services			12.4	

Source: Financial Times R&D Scoreboard, FT Director, 19 September 2000.

example is Vodafone, a 'single play' company focused only on mobile. Other examples are Qwest in the US and COLT in the UK that from the outset were focused to a far greater extent than the incumbents on Internet and corporate data traffic. However, the market capitalization of the latter companies declines as the generalized bull market, after an unprecedented rise, turns into a bear market.

The Co-evolution of R&D

How does R&D co-evolve in the four periods with the changes taking place in companies and the industry as a whole?

Period 1. In Period 1, the pre-liberalization period when the telecoms companies were government-controlled monopolies, corporate R&D was organized in central research laboratories that had an arm's-length relationship with the company's businesses. The classic blueprint for the organization of corporate R&D in the Telecoms Industry (widely copied in other industries) was published in 1971 by Dr Jack A. Morton, who became vice president of Bell Laboratories in charge of electronics technology. Morton's book was titled, *Organising for Innovation: A Systems Approach to Technical Management*. The ability of Bell Labs, not only to create technologies for AT&T and the global Telecoms Industry, but also to win eventually a total of seven Nobel Prizes, became legendary.

In his book Morton went out of his way to emphasize the importance of separating the R&D function in a centralized research laboratory such as Bell Labs. In this way, he argued, the longer-term and more fundamental focus of R&D could be protected from the vicissitudes and demands of the businesses which had to concentrate more on the task of meeting current market demands on the basis of current technologies. Morton was very aware of the dangers inherent in such separation. He knew that R&D could become something of an ivory tower with R&D engineers in a separated central research laboratory (often located in rather idyllic 'green' surroundings) losing touch with the demands of the businesses. It is precisely here where his systems engineering approach came in handy. Conceptualizing innovation as a system that involved the components of research, development, manufacture, and marketing, Morton elaborated on the forms of human communication, and the ways of organizing such communication, that would allow R&D to link effectively with the rest of the company's activities while retaining its semi-autonomous status.[1]

This 'classical' form of organizing R&D in the Telecoms Industry outlined by Morton, and replicated with greater or less modification in the other major telecoms incumbents, was the outgrowth, it must be stressed, of a historically-specific set of circumstances. The state-controlled monopoly telecoms companies that established and modified these central research laboratories were very well aware that innovation was an extremely important means to achieve their goals. Although they did not face competition in their service markets, they did face considerable pressure, reinforced through the political process, to increase both the quantity and the quality of their services. In the early days there was great pressure to reduce waiting lists for telephones and users continually demanded better and cheaper services. Furthermore, there was also strong competition between the major monopoly companies—such as AT&T, BT, FT, and NTT—to be the first or the fastest. While this competition was tempered by cooperative sharing of information and technology—such as took place during the international switching conferences—the rivalry beneath the surface remained strong.

---

[1] Morton, J. A., 1971, *Organising for Innovation: A Systems Approach to Technical Management*, McGraw-Hill, New York. For an account of Morton's influence in Japan see Fransman, M., 1995, *Japan's Computer and Communications Industry*.

However, the absence of strong commercial pressure, exerted by aggressive rivals in the service markets, meant that R&D could more easily achieve the purpose of looking toward the future, to the next generation of network technologies and equipment. This regime, it must be emphasized, was highly innovative. This is proved by the substantial, radical, and cumulative advances that were made by monopoly telecoms companies that included the transistor, contributions to the laser, digital switching, and cellular communications, to name but a few. But the absence of strong commercial pressure did give extra degrees of freedom to the leaders of R&D, both in terms of resources and the ability to organize their activity.

During Period 1 the R&D Conundrum was muted. Since the telecoms monopolies operated on a 'cost plus' basis, and R&D was regarded as one of the costs, there was relatively little pressure on the R&D function to prove that it was paying its way. As we shall see, in later periods this fairly happy situation was to change dramatically.

Period 2. This degree of freedom of R&D leaders began to be whittled down in Period 2. During this period, as is shown in Table 8.1, new entry and competition began to become increasingly important. The competitive performance of the telecoms company's businesses became a priority. Illustrative of the new conditions, Bob Allen, the new CEO of AT&T (which had been separated from the seven regional Baby Bells in 1984) divided the company in 1988 into some twenty businesses grouped into divisions. This reorganization—aimed at increasing the focus, flexibility, and incentive of the businesses—impacted crucially on the organization of Bell Labs.

It was left to Arno Penzias, Vice President of Research at Bell Labs, to deal with the implications.[2] Penzias was in charge of Area 11 of Bell Labs, the part responsible for research. Area 11 employed approximately 10 per cent of the total staff of Bell Labs, the remaining employees being involved with development. Most of the development people in what was still called Bell Labs became the organizational responsibility of AT&T's various businesses.

The most pressing question for Penzias, given the increasing competitive pressure that AT&T was facing, was how the research function in Bell Labs could help the businesses to become more competitive. Penzias's solution, significantly, was to maintain the independence of Bell Labs while making it more responsive to the immediate needs of the businesses. Independence was maintained by ensuring that Bell Labs continued to get the vast majority of its budget paid for from central corporate resources. This, he argued, allowed Bell Labs to remain relatively independent of the businesses' immediate needs so as to be able to look to the longer term.

In order to make the laboratories more responsive to the requirements of the businesses Penzias introduced a major innovation, as important for the reorganization of Bell Labs as it was simple. All the research groups within the

---

[2] This section is based on interviews in Bell Labs with Arno Penzias and his colleagues.

laboratories were required to define customers for their output. In turn, interaction was required with these customers to ensure that the research of the researchers had relevance for the customers. In most cases the customers for the Bell Labs researchers were located in one or more of AT&T's businesses. Other innovations involved new forms of interaction between researchers and their customers in the businesses such as joint R&D projects.

But what about basic researchers? Could the 'customer concept' be applied to them? The answer to these questions was that the customers for basic research were in reality applied researchers. Accordingly, basic researchers were required to define their applied research 'customers' and interact with them in order to try and meet their requirements.

Responding to the same set of competitive pressures, Alan Rudge in BT made similar attempts to get the R&D staff of the central research laboratories at Martlesham to become more responsive to the needs of the company's businesses. Rudge developed what he called 'the customer–supplier principle' as his basic organizing principle for research. While this principle had similarities with Penzias's notion of customer-driven research, Rudge went one step further: 'To ensure that... [research] is well coupled to the needs of the business, the strategy demands that at least two-thirds of the work performed by the central laboratories should be sponsored and financed directly by the customers in the operating divisions [of BT].'[3] Like in AT&T, development people were tied more closely to the businesses.

Like AT&T and BT, NTT faced similar pressures to link R&D more closely with the needs of the company's businesses. However, just as AT&T decided to maintain the corporate funding of the central research laboratories—thus resisting Rudge's decision to make the labs more dependent financially on the businesses—so Dr Toda at NTT decided to do similarly. However, NTT's laboratories were at the same time reorganized and again more of the development staff were organizationally incorporated into the businesses.[4]

During Period 2, therefore, as the responses of AT&T, BT, and NTT make clear, the R&D Conundrum became more important with the central research laboratories of the incumbents reorganizing themselves in order to ensure that they became more responsive to the immediate needs of the businesses.

New entrants and R&D. How did the new entrants who entered the Telecoms Industry in Period 2—such as DDI and Japan Telecom in Japan, Mercury in the UK, and MCI and Sprint in the US—behave in terms of R&D? Significantly, their response was to try and avoid R&D and outsource it to specialist telecoms equipment companies. This choice had the advantage of keeping entry costs lower than what they would have been with internal R&D and making the new entrants simpler organizationally than their incumbent counterparts. However,

---

[3] Quoted in Fransman, M., 1999, *Visions of Innovation*, Oxford University Press, p. 123.

[4] A more detailed analysis of the responses of AT&T, BT, and NTT in the early 1990s is to be found in chapter 3 of Fransman, M., *Visions of Innovation.*

as we shall see later, the decision to completely outsource R&D does raise the question of whether this is an efficient solution in the longer run.

Period 3. Period 3, from the mid-1990s to 2000, saw a continuation of the patterns that were established in Period 2. As shown in Table 8.1, the incumbent telecoms operators continued to do research in their central research laboratories that tended, however, to become significantly smaller largely as a result of the effective transfer of development people to the control of the businesses. Under growing competitive pressures there was also a tendency to reduce long-term and basic research (although this is hard to quantify and to the present author's knowledge there are no studies that have been done on the extent of this important phenomenon).

During Period 3 there was a further significant change. This was the increasing tendency on the part of the incumbents to leave network-related R&D to specialist technology suppliers. The latter included companies like NEC and Fujitsu in Japan and Ericsson, Alcatel, and Siemens in Europe. In line with this tendency in September 1995 AT&T made the decision to spin-off its internal telecoms equipment division that became Lucent Technologies. Henceforth, the central research laboratories of the incumbents focused increasingly on the development of new services leaving traditional telecoms equipment and new data networking equipment R&D to the specialist suppliers.

Pressure from the R&D Conundrum continued during Period 3 as the central research laboratories of the incumbents remained under pressure to prove their worth to their business customers. However, during this period the new entrants continued to rely on outsourcing R&D rather than building their own R&D facilities. As part of the outsourcing process, close relationships developed between the new entrant and its main technology suppliers.

Period 4. Period 4 saw radical changes—and radical threats—facing central research activities in the incumbent telecoms companies. The analysis of these changes is a central theme in the present chapter. However, in order to understand the changes it is necessary to do two things. The first is to examine how the location of R&D in the Telecoms Industry has changed during the four periods. The second is to understand more about how some of the incumbent telecoms companies themselves have changed during Period 4 in response to the pressures of competition and the changing fundamentals in the industry. These two issues are tackled in the following two sections.

THE CHANGING LOCATION OF R&D IN THE TELECOMS INDUSTRY

The changes that occurred in the location of R&D by 2000 are shown in Table 8.2.

Two important points emerge from Table 8.2. The first is the extent to which R&D-intensive activities have moved from the incumbent telecoms companies to the specialist telecoms suppliers. This is true in absolute terms (in terms of total R&D spend), in terms of R&D intensity (R&D as a percentage of sales), and in terms of R&D per employee. (NTT remains the most R&D-committed of the 'Big Five' incumbents. However, its R&D intensity is far below that of the specialist telecoms suppliers as is its R&D per employee. Even in absolute terms its R&D spend is not much different from many of the specialist telecoms suppliers.)

The second (rather dramatic) point to emerge from Table 8.2 is that the telecoms sector as a whole (that excludes telecoms equipment suppliers) is not particularly R&D-intensive when compared to other sectors. Indeed, sectors that are not normally thought of as 'high-tech'—such as automobiles, beverages, and personal care—have R&D intensities that are the same or higher than the telecoms companies.

These two points have several important implications for the evolution of R&D in the Telecoms Industry. Most significantly, they reveal how the process of vertical specialization has produced specialist telecoms equipment and technology suppliers who are capable of providing most of the equipment and technology required by the telecoms companies that operate networks and provide telecoms services. It is precisely for this reason that the new entrants have been able to successfully enter the Telecoms Industry without having substantial internal R&D capabilities, something that would be impossible in other industries such as semiconductors or pharmaceuticals.[5] This has meant that all telecoms companies—both incumbents and new entrants—are able increasingly to leave large parts of their equipment and technology requirements to the specialist telecoms suppliers. In turn this implies a steadily shifting set of R&D priorities for the R&D laboratories of those telecoms network operators and service providers that do their own R&D.

THE REORGANIZATION OF INCUMBENT NETWORK OPERATORS

As shown in detail elsewhere,[6] in response to the radically changed conditions in Period 4 (summarized in Table 8.1), AT&T, BT, and NTT undertook significant reorganizations in 1999/2000. The essence of these reorganizations is summarized in Fig. 8.1.[7]

---

[5] Excluding generics pharmaceuticals companies which 'piggy back' on the R&D of the main pharmaceuticals companies. [6] See Chapter 3.

[7] Not all the 'Big Five incumbents' have undergone such radical reorganizations. Deutsche Telekom and FT by 2000 had not made such significant changes as AT&T, BT, and NTT to their corporate organizational structure. See Chapters 4 and 5.

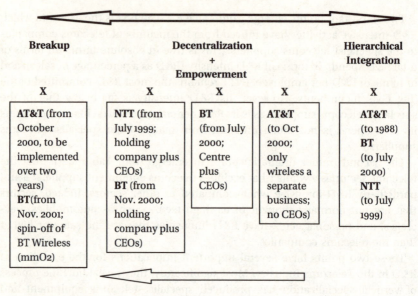

Fig. 8.1. The evolution of corporate organization in AT&T, BT, and NTT.[8]

Source: M. Fransman.

The main point to emerge from Fig. 8.1 is that AT&T, BT, and NTT—the first companies in the large industrialized countries to face liberalization and competition—have all tended to move in the same direction in terms of their corporate reorganization. This is shown by the light arrow at the bottom of the diagram. More specifically, all three companies have moved away from 'hierarchical integration' where they integrated their various activities under the hierarchical control of the company's 'Centre'. Instead they opted for 'decentralization and empowerment'. Under this form of corporate organization they segmented their activities into semi-autonomous businesses that were given significantly more independence from the Centre. In AT&T's case, the decision was made on 25 October 2000 to break the company up into four independent companies.

The reorganization of AT&T and BT has posed significant challenges, indeed threats, for the R&D function in these companies. The main reason is that the R&D Conundrum is confronted in a particularly acute form under the new organization. More specifically, a new dimension has been added to the conundrum. Not only is there the question of why company resources should be allocated to R&D when benefits are uncertain and only accrue in the future. In addition, further questions are raised regarding the costs and benefits of central research for the independent or semi-independent businesses. Very concretely the heads of the businesses frequently ask: 'Why should we be contributing to central laboratory research when we have limited control over what research

---

[8] This diagram is also analysed in Chapter 3.

they will be doing and when the benefits of the research have to be shared with the other businesses?' Clearly, these further questions pose significant challenges to R&D leaders.

How have BT and AT&T negotiated and attempted to resolve the R&D Conundrum in Period 4? In the following subsection the response of BT is analysed while in the subsequent subsection the preliminary response of AT&T is examined.

BT's R&D Response[9]

The location of the Technology Function in the new, reorganized BT is shown in Fig. 8.2.

As can be seen from Fig. 8.2, the Technology Function is one of the responsibilities of the 'Centre' of BT, together with the Executive Committee and the Investment Committee, both under the CEO. Furthermore, the Technology Function has two components, Group Technology (BT's terminology) and (in my terminology) Technology Resources. Hitherto, these two were joined together in Martlesham Laboratories and were responsible to the CTO who was a Board Member, namely Dr Alan Rudge.[10]

Of these two organizations it is Group Technology that is the more innovative, representing a significant break with the past. Technology Resources consists of approximately 3,500 R&D people who were part of the Martlesham Laboratories, now called Adastral Park. They continue to perform much the same function as they did before the reorganization, namely doing the R&D needed by BT's seven new businesses. (In quantitative terms—both in terms of people and budget—most of them are involved in development activities, closely related to the needs of the businesses.)

Group Technology contains only about eighty people. However, it is the function that they perform that constitutes the significant innovation. Essentially, Group Technology has three closely-related aims, that involve *looking longer term* to identify:

- Paradigm-changing technologies
- New service propositions
- New business areas.

Significantly, Group Technology does not see the *acquisition* of paradigm-changing technologies as necessarily requiring that BT itself should get or

---

[9] This section is based on discussions with Dr Tom Rowbotham, head of BT's Group Technology, and several of his colleagues who are not responsible for the analysis here of BT's R&D.

[10] For details on the organization of R&D in BT under the company's old regime, and for a comparison with AT&T and NTT, see Fransman, M., 1999, *Visions of Innovation: the Firm and Japan*, Oxford University Press.

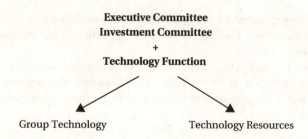

Fig. 8.2. The technology function in the reorganized BT.

develop the technologies concerned. In some cases it may be sufficient that the company ensure *access* to the technologies. For example, BT sold one of its R&D units dealing with optical communications devices to Corning (the manufacturer of optical fibre) that is co-located with BT at Adastral Park. The agreement with Corning requires that BT be given access to the technology and the subsequent improvements made to it over time. BT and Corning continue to share information in this area. In practice, Corning works with the telecoms equipment makers, Nortel and Lucent, who incorporate the technology into their equipment, which is then sold to BT. In this way BT gets the fruits of the technology without actually having to acquire the technology (either by buying it or internally developing it). BT's plan is that the Adastral Park site will increasingly be a multi-company site and knowledge will flow between the company R&D organizations in the park. For instance, Nortel (that has been chosen as the main supplier of third-generation mobile equipment to BT Cellnet/mmO2) is also co-located in Adastral Park.

The second of Group Technology's aims is to develop 'new service propositions'. This allows BT to benefit from the longer-term perspective and sometimes different 'mindset' that R&D workers often have. Unlike the employees of the businesses who of necessity are tied to *existing* markets and technologies, the R&D people may be able to develop a different, innovative perspective (as is clear from the experiences of highly innovative R&D organizations such as Xerox's Palo Alto Research Center, PARC). One source of new service propositions is the 3,500 R&D staff in Technology Resources who are invited to suggest potential new service propositions. Group Technology then selects from among the suggestions.

The following quotation from the author's interview with Dr Rowbotham captures well the innovative spirit of Group Technology and its break with the traditional R&D activities of the telecoms incumbents:

The budget for Group Technology consists largely of the 'R' component of R&D and is about 10 per cent of the total R&D budget. We can do what we like with it. *We don't even have to spend it on technology. I see myself as a business angel.* Many of my people work closely with the corporate strategy people in the Centre and in my group the technology function and the corporate strategy function come closely together. There also is a new business incubator on the site with small venture start-ups beginning with as few as three people, 'facing' different venture capitalists and partners (emphasis added).

Group Technology is also a 'customer' for Technology Resources when it commissions work from the latter.

Brightstar. On 19 November 2000 BT established a new division, Brightstar, a technology incubator aimed at setting up dozens of new technology-based businesses that will be sold or floated on the stock market. BT's hope is that Brightstar will be worth £2 billion over three years. The intention is that Brightstar will invest in up to ten new ventures at any one time. Each venture would spend up to four months in the incubator where it will receive accommodation, managerial, and financial support. Brightstar will take minority holdings in the start-ups which will use BT's patents and intellectual property. BT currently owns around 14,000 patents.

R&D and BT's Businesses. Each of BT's seven businesses has three options when they require R&D and the acquisition of technology:

1. Get technology from the Centre (i.e. from Technology Resources and/or Group Technology).
2. Get it from outside BT.
3. Do it themselves (i.e. in the business).

Each business has its own R&D facilities (that concentrate primarily on 'D'). Furthermore, each business has its own CTO. However, the power of the business CTOs is constrained by the Centre. This constraint works in the following way. The head technology officer for BT as a whole is a 'CTO of CTOs'. He sits on both the Executive Committee and the Investment Committee (see Fig. 8.2). In this capacity he has the ability to turn down R&D investment proposals that come from the businesses. He may reject an R&D investment proposal from a business for various reasons. For example, he may feel it does not make sense for BT as a whole; he may feel that it involves unnecessary duplication of R&D being done elsewhere in the company; or he may simply feel that it does not make sense at the level of the business. Therefore, although each business has the three options outlined in terms of R&D, the businesses remain constrained by the Centre.

Conclusion. BT's reorganization has also involved a significant change in the Technology Function in the company (although it must be added that this change has not come overnight, but has been evolving since the reforms in the R&D area brought about by Dr Alan Rudge who emphasized the role of the customer—both external to BT as well as within the company—in orienting the direction of R&D).[11] Overall, the mindset in BT has changed significantly from

[11] See Fransman, M., 1999, *Visions of Innovation*, Oxford University Press, for further details.

the days of the classical incumbent central research laboratory. Instead of a focus on technology *per se*, the focus of those performing the technology function is now on:

1. BT's strategy.
2. Paradigm-shifting changes and new service propositions.
3. Technology acquisition and exploitation, that may require, *not* R&D done internally, but, for example:
   - spinning a new company out of BT;
   - *selling* technology to another company;
   - or getting partner companies to provide the technology.

These significant changes in mindset regarding the technology function are captured in Tom Rowbotham's words: 'I see myself as a business angel' and 'We don't even have to spend our budget on technology'!

It is clear from this analysis of the recent changes in BT's R&D organization that the R&D Conundrum has been negotiated and resolved in a very different way compared to earlier periods.

AT&T's R&D Response[12]

The impact of AT&T's breakup into four separate companies on the research function of AT&T Laboratories is shown in Fig. 8.3.

As shown in Fig. 8.3, the Research Division of AT&T Laboratories is to be moved into one of the four companies into which AT&T will be broken up (announced on 25 October 2000), namely AT&T Business Services. (The Research Division descends from Area 11 (Research), under Dr Arno Penzias, in the old Bell Laboratories that existed before AT&T voluntarily trivested itself in September 1995, spinning off Lucent and demerging NCR.) *In effect, this means that AT&T is abandoning a central research laboratory organized separately from the company's businesses.* However, Lucent continues to have its central research laboratory, consisting of the bulk of R&D workers who were in the old Bell Laboratories.

How will the Research Division be organized within AT&T Business Services? At the time of writing the answer to this question is not entirely clear. However, one possibility is that the Research Division will become a separately incorporated subsidiary of AT&T Business Services. Funding has been guaranteed to the Research Division for a period of three years. After this time the Research Division must become self-sufficient. It is expected that further funding will come from all of the four AT&T companies. An important further provision is

---

[12] This section is based on information provided to the author by senior AT&T researchers.

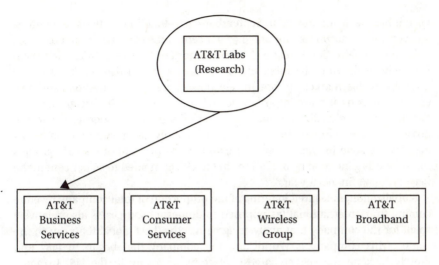

Fig. 8.3. The R&D function in the post-breakup AT&T.

that the Research Division has been given complete control over AT&T's considerable patent and technology portfolio and it is hoped that this resource will also make an important contribution to future funding.

Clearly, these changes imply a major rupture with the past. Most importantly, at least in the first instance, the changes mean that three of AT&T's compan-ies—AT&T Consumer Services, AT&T Wireless Group, and AT&T Broadband—will from an organizational point of view much more closely resemble their new entrant counterparts. More specifically, they will outsource their R&D requirements, either to the Research Division in AT&T Business Services, or to some other source such as specialist technology suppliers. It remains to be seen how the Research Division's role will evolve in the future under AT&T Business Services.

In the next section but one we reflect on the implications of the changes in the R&D function in BT and AT&T.

GLOBALIZATION AND R&D

How important is globalization in the Telecoms Industry as the twenty-first century dawns? The answer is not very important. Globalization, at this time in the industry's evolution, is more rhetoric than reality.

In 1997 AT&T reiterated the importance of globalization as one component in its 'three part strategy', emphasizing its goal to 'broaden and deepen our global reach'.[13] However, in 1999 AT&T reported—after the de-merger of NCR and

---

[13] AT&T Midyear Report, 30 June 1997, p. 1.

Lucent had been announced in September 1995—that 'international operations and ventures'[14] accounted for only 1.97 per cent of the company's total revenue.

For BT, globalization is relatively more important than for AT&T. The reason is largely to do with the relatively small size of the British market and the invasion of this market by foreign telecoms operators and by domestic new entrants, making it more important for BT to seek growth in foreign markets. For this reason globalization has been a priority for BT as soon as it became possible for the company to invest in other (mainly European) markets. However, by the 2000 financial year 'approximately 95 per cent of the [BT] group's turnover was generated by operations in the UK, compared to 96 per cent in the 1999 and 1998 financial years'.[15]

Globalization, measured in terms of the proportion of total company revenue coming from international (as opposed to domestic) sources, is also relatively small for the company that analysts agree has the 'best global footprint in the world'. This company is WorldCom, the company that began in 1983 and rapidly became the second largest telecoms company in the US. In 2000, however, only 5 per cent of WorldCom's revenue came from international sources.

With this low degree of globalization of revenues it is hardly surprising that the globalization of R&D in the Telecoms Industry has also been slight. AT&T does have a small laboratory in Cambridge, England, run by Professor Andy Hopper. BT has several small R&D operations in various parts of the world, none of which plays a significant role in the company's overall R&D. NTT has a small laboratory in Silicon Valley, although in 2000 NTT DoCoMo announced a more substantial laboratory that will be established in Munich, Germany to examine fourth-generation mobile communications.

It is clear, therefore, that the globalization of R&D in the Telecoms Industry is, at best, at an early stage in its evolution. However, it has to be said that the tendency for the incumbent telecoms companies—the only network operators that are doing any significant R&D—to segment their business activities and even to break their companies up completely does not augur well for the future of globalized R&D in the Telecoms Industry. The smaller, more focused companies and businesses that are being established by the incumbents are likely to have a smaller commitment to globalized R&D than their former larger parents. However, the example of NTT DoCoMo serves as a reminder that even smaller, highly-focused telecoms operators may still see benefit in globalized R&D. Vodafone, not very R&D-oriented itself, has inherited an R&D laboratory in California.

---

[14] 'International operations and ventures' include: AT&T's consolidated foreign operations, international carrier services, international online services, and the earnings or losses of AT&T's non-consolidated international joint ventures and alliances. However, bilateral international long-distance traffic is excluded. AT&T Annual Report, 1999, p. 24.

[15] BT, Annual Report, 2000, p. 39.

REFLECTIONS

Period 4 (continued)

What are the implications of the changes in the role and organization of R&D in AT&T and BT for the broader issue of the co-evolution of companies and R&D in the Telecoms Industry in Period 4? The most important implication is that corporate R&D in the network operator layer of the Telecoms Industry is rapidly losing its 'protected' status. This has extremely important implications, not only for individual R&D-performing companies, but also for the Infocommunications Industry as a whole as will be shown in this section.

In what sense has corporate R&D in this layer of the industry been 'protected' in incumbent companies such as AT&T, BT, FT, and NTT? Corporate R&D in these companies has until recently been protected in the sense that the heads of these hitherto hierarchically integrated companies have accepted the notion that R&D is necessary for both the short-run and the long-run competitiveness of the company. As a result, a significant proportion of the R&D budget has been funded from corporate sources. As we saw earlier, even Alan Rudge, writing about BT's R&D philosophy in 1990, acknowledged that one third of the R&D budget of the central research laboratories should come from corporate sources. And both AT&T and NTT went much further than this, insisting that the vast majority of the budget was corporately funded.

The effect of corporate funding was to give the central research laboratories a reasonably certain income stream, even though the absolute sum was subject to the normal negotiation and horse-trading. Getting corporate resources meant the laboratories could look beyond the immediate needs of the businesses, based as they inevitably are on current markets and current technologies. Longer-term and more fundamental research and the next generation of technologies accordingly became important. In this sense the laboratories were to some extent protected although, of course, they still had to convince the sometimes sceptical leaders of the company that they were paying their way by adding both short-term as well as longer-term value to the company.

As shown in Table 8.1, however, in Periods 2 and 3—from the mid-1980s to 2000—this protected status has gradually been whittled away. Increasingly, the R&D Conundrum has been resolved in ways that have involved making the company's businesses the drivers of the R&D function. Typically in these two periods this has involved insisting that more and more of the budget of the central research laboratories comes from contracted R&D negotiated with the businesses. With the businesses as the paying customers of the central research laboratories, and their payments constituting the bulk of the laboratories' budget, *current* business needs have increasingly come to dominate the focus, priorities, and activities of the laboratories. In this sense their protection has diminished and their ability to look beyond the immediate needs of the businesses reduced.

With the reorganization of AT&T and BT involving decentralization plus empowerment and breakup—shown in Figure 8.1—not only has the tendency for the whittling away of protection been accelerated, the whole role of R&D has been transformed. Not only are the central research laboratories seen as organizations that must be responsive to the current needs of the businesses, *they are seen as organizations that must themselves become businesses*. This is seen very clearly with the requirement that the Research Division of AT&T Laboratories must become self-sufficient within three years after the breakup of AT&T. It is also seen in the establishment of BT's Brightstar that is intended to facilitate the direct generation of revenue and shareholder value from the R&D function.

At this point in the analysis, two questions need to be posed. First, what forces have brought about this fundamental change in the role of R&D in AT&T and BT? Second, is the change good or not?

The answer to the first question is that the change is the logical outcome of the evolution of corporate organization in these two companies from hierarchical integration (where central R&D was 'protected') towards decentralization plus empowerment and breakup. The reason is that the reorganization of the companies has meant that the companies have been segmented into more autonomous businesses that are far more focused on competition and profit. However, both competition and profit are driven by short-term considerations. Under these circumstances the R&D Conundrum becomes acute. And the resolution of this conundrum has involved forcing the central research laboratories to focus more of their attention and resources to meeting the immediate needs of the businesses. In the case of AT&T this process has gone so far that the central laboratories themselves have been abandoned and moved into one of the businesses.

But is this trend good or bad? The answer is that the trend is good *and* bad. It is good in so far as R&D is more tightly coupled to the needs of the businesses and therefore hopefully helps the businesses to become more competitive and profitable. To the extent that this is indeed the outcome, the businesses have every reason to strengthen their commitment to R&D. However, it is bad because in this process the role and objectives of corporate R&D have been transformed. Although corporate R&D has always had as one of its objectives the meeting of the needs of the company's businesses (even in so-called fundamental or basic corporate R&D), this has only been one objective, even if the most important. Another important objective has been to look to the future, and better prepare the company to survive and prosper in the future, by deepening knowledge of existing products, processes and materials, and increasing knowledge of possible future technologies.

The danger is that in Period 4, under the pressures induced by liberalization, the role of R&D in the network operator layer of the industry will be twisted too far in the direction of meeting the immediate needs of the company's businesses. Does this matter? The answer is yes, if one believes in the value generated by R&D focused towards the longer term. And it is well known that such R&D has yielded extremely rich fruit in the past, as examples such as Bell Laboratories and Xerox PARC clearly testify.

The problem, however, is that the example of Xerox PARC also clearly illustrates that a distinction needs to be drawn between benefits that are appropriable by the company financing, sponsoring, and housing the R&D, and benefits that can also be appropriated by others. It is well known, for instance, that some of Xerox PARC's most important innovations were appropriated by other companies, such as Apple, rather than Xerox.

This points to an important dilemma, the resolution of which will have to go far beyond the companies that fund and undertake the R&D: R&D that is oriented towards the longer-term future can provide significant benefits for the Infocommunications Industry and for society as a whole; but the pressures and incentives that are being put on the R&D-performing companies in the network layer of the Telecoms Industry are tending to reduce the priority that is being given to such R&D. It seems clear that this is a key issue for global public policy.

New Entrants in Period 4. For new entrants, such as WorldCom, Qwest, and Level 3 in the US and COLT and Energis in the UK, the R&D Conundrum, as already noted, has not been so much of an issue. The reason is that these companies have chosen on the whole to outsource their R&D rather than do it themselves, as is shown in Table 8.1. But even they are discovering that as their operations become bigger and more complex, and as their networks and technologies become more complicated, so it is necessary to have more in-house knowledge. At the moment there is little sign that this requirement for increased in-house knowledge is leading the new entrants to establish specialized R&D laboratories, located either within businesses or in central facilities. But at the dawn of the twenty-first century it is not yet clear whether the new entrants will eventually see the need to grow their own substantial R&D competencies.

CONCLUSION

In this paper it has been shown that the evolution of industries, companies, and the R&D function is a co-evolutionary process. The evolution of the Telecoms Industry has been divided into four periods (see Table 8.1). During these four periods the role of R&D, and of central R&D laboratories, has changed in important ways. An important driver of change has been liberalization and the competitive pressures that it has created. This has put more pressure on the incumbent telecoms companies to enable their businesses to become more focused, flexible, fast, and flat—just like their smaller, agile new entrant competitors. In turn this has meant that the R&D Conundrum has become more acute. This conundrum arises because some of the leaders in a company sometimes do not see why R&D should be generously funded since its output is uncertain and only accrues in the future.

The attempt to resolve this conundrum, as has been shown in detail in this chapter, has taken the form of making the company's businesses drive its R&D. However, while this has had the positive effect of making the R&D laboratories more responsive to the immediate needs of the businesses, it has also had the negative effect of decreasing the priority accorded to longer-term R&D. This may have negative consequences, both for the company itself as well as for the Infocommunications Industry and society as a whole.

To put it bluntly, longer-term research carried out in 'classical' central research laboratories, such as the old Bell Laboratories, has on occasion led to path-breaking new technologies, the transistor being a key example. Where will this kind of research come from in future? Will it matter if it no longer takes place? These are key questions.

To conclude, R&D in general, and central laboratories R&D in particular, have potentially a major contribution to make to companies' competitiveness and profitability. However, the 'R&D Weapon' has constantly to be remoulded as the industry and its companies evolve. This paper has shown how the R&D function in the network operator layer of the Telecoms Industry has changed as this industry and its companies have evolved. However, R&D itself remains, and will remain, one of the most important enablers of longer-run survival and prosperity in the Infocommunications Industry. The question is how to ensure, as competitive pressures grow and consolidation takes place, that there are sufficient incentives to undertake R&D, a major engine of longer-term industrial growth and prosperity.

9

Mobile Internet: Explaining Japanese Success and European Failure

In this final chapter of *Telecoms in the Internet Age* we examine the emergence of the mobile Internet. The mobile Internet allows the Internet to be accessed from mobile devices such as mobile phones and handheld computers.

The mobile Internet first emerged in the late 1990s when parallel attempts were made in Japan and Europe to reformat Internet web pages so that they could be accessed from mobile phones. These attempts present an important puzzle. The puzzle arises since the Japanese attempt resulted in resounding success while the European efforts were a dismal failure (though further efforts are being made to improve the European outcome). Why was there such a difference in outcome in these two regions of the global Telecoms Industry? This is the main question that is tackled in this chapter.

INTRODUCTION

In the chapters on the Big Five incumbents—AT&T, BT, Deutsche Telekom, France Telecom, and NTT (Chapters 3–5)—it was shown that mobile telecoms was their most important source of revenue and profit growth during the late 1990s and into the new millennium. Mobile voice services compensated for their slowing performance in their core business of fixed long-distance and international voice services. Vodafone was the most important beneficiary from the spectacular growth in mobile voice in the 1990s. As a 'single-play' mobile

I would like to acknowledge interviews with several NTT DoCoMo people, including Dr Keiji Tachikawa, President and CEO, Mr Kiyoyuki Tsujimura, Senior Vice-President, and Mrs Mari Matsunaga, a member of the original i-mode development team. However, neither the content nor the interpretations in this chapter should be attributed to them. This chapter draws heavily, particularly in analysing mobile Internet developments in Europe, on the excellent paper by Jon Sigurdson, which provides the first detailed account of the development of WAP (wireless application protocol). The paper, 'WAP OFF—Origin, Failure and Future', can be downloaded from http://www.TelecomVisions.com.

company, Vodafone was not dragged down by the slowing growth in revenue and profits from fixed long-distance and international voice services.

However, by the early years of the new millennium the mobile Telecoms Industry was at a crucial watershed in its evolution. The reason was the growing saturation of mobile voice services in the rich countries. By 2002 this saturation appeared in the form of slowing sales of mobile phones.[1] Furthermore, the slower than previously anticipated transition to later generations of mobile networks (2.5- and third-generation) added to the slowdown in the entire mobile industry, affecting both mobile operators and mobile equipment suppliers.[2]

By the mid-1990s far-sighted mobile operators were already beginning to worry about what would replace mobile voice as a source of growth for the mobile industry. In Japan, for example, Mr Ohboshi, President of NTT's DoCoMo mobile subsidiary that was organizationally separated from NTT in 1992, began to take steps from around 1996 to deal with this problem. It was clear to him that just as was happening in the area of fixed communications, mobile would also have to make the transition from reliance on voice services to the increasing provision of data-based services. The big question was what would these data-based services be and how would they be delivered.

In Europe, significantly, the thrust came, not from the European mobile operators, but from the main globally dominant mobile equipment suppliers, Ericsson and Nokia.[3] In the US, a small mobile startup, Unwired Planet (later renamed Openwave), and Motorola, the main American mobile equipment maker, also became involved with the Europeans. These companies, at first individually but then collectively, attempted to develop new standards for reformatting web pages so that they could be viewed over mobile phones.[4] Eventually this led to the announcement in the spring of 1998 of the establishment of the wireless application protocol (WAP) Forum which was formally set up in December of that year.

However, by 2002 WAP had proved to be a dismal failure. By the end of 2001 there were approximately 640 million GSM[5] mobile subscribers in the world. About one-tenth of these used Internet-enabled WAP phones. Hardly any of

[1] In the developing countries, however, including Russia and Eastern Europe, growth rates are still rapid. By 2001 China was estimated to have 125 million mobile subscribers, making it the largest market in the world, rising to around 350 million by 2005.

[2] Although not done here, the mobile industry can also be usefully examined using a similar Layer Model to that developed in Chapters 1 and 2. By April 2002, this slowdown even had a significant negative impact on Vodafone's share price.

[3] The reason for this crucial difference is analysed later.

[4] Sigurdson, op. cit., provides a detailed account of the standards at first developed inside these companies and the conflicts and negotiations that took place as they eventually agreed on a common approach.

[5] Global System for Mobile Communication (GSM) was the European second-generation, digital, mobile standard that emerged originally from pre-existing Scandinavian standards. The European mobile equipment suppliers, mobile operators, and the European Telecommunications Standardization Institute (ETSI) worked together in order to establish GSM as a European standard. It was later widely adopted outside Europe and has formed the basis for subsequent

them used their phones to access the Internet. In Japan, however, Internet-enabled mobile phones took off as soon as they were introduced by the leader and still-dominant company, NTT DoCoMo, that introduced its branded mobile Internet service called i-mode in February 1999. By 2002 i-mode had over 30 million subscribers and this service generated about 15 per cent of DoCoMo's revenues.

The Puzzle

The Japanese success in stark contrast to the European failure presents a puzzle. Why was there such a difference in outcome in these two regions of the global Telecoms Industry, both apparently trying to do the same thing, namely to Internet-enable mobile phones? This is the main question that is tackled in this chapter.

Contents of this Chapter

In the following section the inadequacies of the conventional 'explanation' of i-mode's success are discussed. Instead an alternative explanation is provided, namely that the success of i-mode is due to the entirely new innovation system that it created. In the next section the properties of the 'i-mode Innovation System' are analysed. But this raises further questions regarding how this system emerged and evolved. These questions are tackled in the subsequent section. The very different evolutionary path in Europe is then examined which resulted in the failure of WAP. Finally, the revenue-generating properties of i-mode and DoCoMo's globalization are discussed.

INADEQUACY OF THE CONVENTIONAL WISDOM ON i-MODE

In their initial attempts to understand the runaway success of the mobile Internet in Japan, and specifically i-mode's success, European explanations tended to focus on the technical characteristics of the Japanese system, which offered users advantages that were unavailable through WAP in Europe, and the initial conditions existing in Japan that differed markedly from those in Europe. Four of the main factors that were emphasized are summarized in Table 9.1.

so-called 2.5-generation technologies such as GPRS (General Package Radio Service), a packet technology for speeding data transmission rates on GSM networks, and EDGE (Enhanced Data for Global Evolution), which allows multimedia and broadband functions to be performed over GSM networks. The third-generation standard, W-CDMA (Wideband Code Division Multiple Access), agreed as a common standard in Europe and Japan and adopted by some mobile operators in the US, has also evolved from GSM.

Table 9.1. Conventional explanations of i-mode's success

i-mode's always-on functionality
bit-based rather than time-based charges
low PC usage and PC-based Internet access in Japan
targeting a mass market of relatively young consumers

DoCoMo's i-mode user had simply to press the i-mode button on the mobile phone in order to reach the menu of i-mode services. Thereafter, the user was continuously connected to Internet-based i-mode services. Furthermore, the i-mode user was charged according to the amount of data sent and received (i.e. according to a bit-based tariff). The WAP user, however, was charged according to the amount of time used in accessing the services (the user experience being much like being in a taxi in a traffic jam with the meter running). As a result the user tended to disconnect after the WAP service was used, but thereafter had to dial up again when a new service was required. All this took a significant amount of time and the user was charged by time, making the service expensive to use.

The fact that Japan had lower adoption rates of personal computers and fixed Internet access than Europe was thought to be another determinant of relative Japanese success. The mobile Internet, it was thought, was relatively more useful for relatively more people in Japan than in Europe. Finally, in Japan a specific market niche, namely young urban consumers, were targeted, whereas in Europe WAP was not specifically targeted.

How adequate is this 'explanation' of i-mode's success? The answer is that these four factors do highlight important differences between Japan and Europe that undoubtedly had an impact on the subsequent diffusion rates of i-mode and WAP services. However, this 'explanation' only scratches the surface and does not get to the heart of the matter. To begin with, a rigorous explanation must go one step back and explain why it was that in Japan technical characteristics were chosen that were clearly advantageous from a user's point of view but were not chosen in Europe. Second, while the four factors shown in Table 9.1 do have a bearing on the performance differential between Japan and Europe, they do not explain the performance differential in Japan between DoCoMo and its two competitors.

The problem with these four explanatory variables, even when they are taken together, is that although, arguably, they were necessary for the success of i-mode, they were not sufficient. This becomes apparent when DoCoMo is compared with its Japanese competitors. DoCoMo now has two main competitors, KDDI and J-Phone.[6] These competitors introduced mobile Internet services in Japan only several months after DoCoMo. Their services offered

---

[6] KDDI was formed in October 2000 as a result of a merger between the mobile operations of DDI and IDO. J-Phone is the mobile subsidiary of Japan Telecom. In 2001 Vodafone took over majority control of Japan Telecom and J-Phone after acquiring BT's share in the companies.

basically the same technical characteristics as DoCoMo's and obviously were introduced under the same Japanese conditions. However, the competitors had a significantly smaller number of subscribers to their mobile Internet services than DoCoMo.[7]

Clearly, therefore, there is more to the DoCoMo i-mode success story than the factors mentioned in Table 9.1. In the following section it will be proposed that the answer to the puzzle lies in the unique innovation system that DoCoMo initially inadvertently, created. The characteristics of this innovation system are analysed while in the subsequent section an account is provided of how this innovation system came to be.

RESOLVING THE PUZZLE: THE 'i-MODE INNOVATION SYSTEM'

It will now be shown that DoCoMo's success with i-mode is to be attributed to the properties of a remarkable innovation system that the company created in developing its i-mode service. This innovation system, it is worth stressing, like most complex socio-economic-technical systems, evolved over time in a piece-meal, iterative way the outcome of which was neither initially intended nor foreseen by its creators. As such, DoCoMo's innovation system bears a close resemblance to another key Japanese emergent innovation system, namely the just-in-time system of production.[8]

The Properties of the 'i-Mode Innovation System'

The properties of what will be called in this chapter the 'i-mode innovation system' (henceforth IIS) are shown in Table 9.2. In the following section a detailed explanation is given of how IIS evolved.

The four properties of IIS will now be considered as a combinant whole.

First Property: Understanding Customer Segment and Tastes

The first property of IIS is an evolving understanding of customer segment and tastes. As is shown in more detail in the following section, the original intention

---

[7] By the end of January 2001, KDDI has about 6 million subscribers to its EZweb mobile Internet service compared to around 25 million for DoCoMo's i-mode.

[8] For an account of the evolution of the just-in-time system see M. Fransman, Knowledge Segmentation-Integration in Theory and in Japanese Companies, in O. Granstrand (ed.), *Economics of Technology*, North Holland, 1994, pp. 170–175.

Table 9.2. The combinant properties of the 'i-mode innovation system' (IIS)

Evolved understanding of *consumer segment and tastes* (i.e. young customers with taste for magazine-like contents accessible while on-the-move)

Creation of *high-powered incentives* for complementary content-creators (i.e. incentives for both DoCoMo and 'i-mode menu sites' to create content)

Technology in the service of evolving consumer demand (i.e. rather than the other way round)

Creating and appropriating the benefits of *network externalities, positive feedback loops, and dynamic increasing returns* (i.e. creating the conditions for cumulative advances)

was for i-mode to become a service for business customers. However, as the i-mode development team contemplated the target market further, it was decided to concentrate instead on the young mass consumer market. This was a key shift in cognitive framework that would have a crucial bearing on the future evolution of IIS. The conditions under which this change in cognitive framework was made are examined in the next section. The point here, however, is that the first property of IIS was a detailed conception, developed over time, on who the consumers for i-mode would be, the kinds of content, applications, and services they would want, and the conditions under which they would find it appealing to consume these services. As will be seen, the 'magazine metaphor' became a key part of the cognitive framework that was used to interpret evolving consumer preferences and tastes.

Second Property: High Powered Incentives for Content Creators

The second property of IIS is high powered incentives for complementary content creators. If the target mass market was to be young consumers the question arose regarding what precisely they would want to consume. The 'magazine metaphor' suggested an answer in the form of the content available to young people in the magazines that they were already purchasing in large quantity. Not only could this kind of content be made easily and cheaply available whilst on-the-move, new forms of innovative content could also be developed, facilitated by the digital mobile information and communications technology (such as the surprise cartoon characters created by a company called Bandai appearing each day on i-mode mobile handsets—discussed later).

But this raised a further problem: How could this kind of content be generated and by whom? The solution to this problem created by IIS constitutes one of the most important elements of the innovation system.

Solving the Incentive Problem: Micro-Payments

It was obviously insufficient for the i-mode development team merely to have ideas about the kind of content that would be suitable for the service, no matter how concrete the ideas were. Equally important, they needed to ensure that the appropriate content was created. Moreover, it was also clear that the content would have to be created by others since, as essentially a mobile network operator, DoCoMo lacked the competencies to become a content creator. (Indeed, even if DoCoMo possessed such competencies it would not be possible for the company alone to provide either the quantity or variety of content that the i-mode service required.)

However, in order to persuade independent content creators to generate magazine-like content for i-mode, it was important that they had a sufficiently strong incentive to do so. It was in this connection that DoCoMo came up with what would prove to be an extremely powerful solution.

Following the 'magazine metaphor' the i-mode team had already decided that the monthly cost of subscribing to the i-mode service should not come to more than the average magazine. Eventually it was agreed that the monthly i-mode charge would be 300 Yen (about $3).[9] In connection with the problem of content creation this raised a number of further questions. If content creators were to be given a sufficient incentive to produce content they would in many cases (though not necessarily all) have to be paid for their content. This payment would have to come from the ultimate consumers of the content. But how much should the consumers be charged for content? How should the money be collected from them? How much should the content creator be paid and how should this payment be made?

It was in answer to the latter questions that the i-mode team came up with its important solution: using DoCoMo's own billing system to collect micro-payments from customers and, after aggregating them, paying them to the content creators minus a charge for this 'billing service'. The team eventually decided that content creators should be allowed to charge customers monthly sums of 100, 200, or 300 Yen and that DoCoMo would deduct 9 per cent as its billing service commission. This left the content creator with 91 per cent of the customer's payment, a high-powered incentive to create content.

In this way DoCoMo was able to create a 'win–win' situation for itself as the mobile operator, for its complementary content creators, and for customers.

---

[9] All charges related to i-mode are outlined later in this paper.

And the cement binding this three-way relationship together consisted of high-powered incentives that gave each of the three parties the motivation to make i-mode a success.

In creating this solution DoCoMo was able to solve a major incentive problem still eluding the Internet, namely the problem of the reward for creating content available on Internet web sites. It is currently still the case that the incentive for those who provide content on web sites on the fixed Internet is severely constrained by the absence of a credible micro-payments mechanism. Essentially, there are only two ways of generating revenue from a regular web site: advertising or through e-commerce transactions. The latter, however, are facilitated by credit card payments that are not suitable for micro-payments of very small amounts. At present the absence of micro-payments constitutes a serious disincentive limiting the creation of content on the fixed Internet. DoCoMo's IIS, however, provides an effective solution to this problem.

Third Property: Technology in the Service of Customers

The third property of IIS is technology in the service of evolving customer demand, rather than the other way round as in the technology-push scenario. Essentially i-mode should be understood as a new set of services geared to the mass consumer market rather than as a new set of technologies. Indeed, the i-mode team tended to choose distinctly 'old' technologies.

For example, although packet-switching was chosen as the transmissions technology (using, as shown in the following section, technology that DoCoMo had already developed), the packets were sent over a network that was partially circuit-switched.[10] This provided the 'always-on' functionality and bit-based tariffing that proved to be exceptionally customer-friendly. When it came to choosing the technology that would be used to i-mode-enable web sites so that they could be viewed over an i-mode handset the team stuck as closely as possible to the existing Internet standards, namely HTML, HTTP, and TCP-IP. Working with other Japanese firms (see following section) they came up with cHTML (compact HTML) that greatly helped content creators by using a computer language with which they were already familiar. (WAP, the corresponding European technology, took somewhat longer to master.) cHTML was developed by a small Japanese firm called Access, that started up in the 1980s.

In short, the entire i-mode operation was very much driven by an evolving conceptualization of what it is that the young target mass consumer market would want rather than by the wizardry that the latest technologies could produce. It is probably in the area of evolving i-mode handsets that technical ingenuity has been most evident in IIS. However, handsets have been developed

---

[10] See Chapter 2 for a detailed discussion of the emergence of packet-switching, one of the key technologies of the Internet, and the way in which it differs from circuit-switching.

in response to evolving consumer demand and here DoCoMo has fortunately been able to rely on an extremely competitive market fuelled by R&D-intensive Japanese electronics companies such as NEC, Mitsubishi Electric, Matsushita, and Sony (although DoCoMo, far more R&D-intensive than other global mobile operators such as Vodafone, has itself been closely involved in handset design).

Fourth Property: Network Externalities, Positive Feedback, and Dynamic Increasing Returns

With the main elements of IIS firmly in place, DoCoMo was able to benefit from significant first-mover advantages. With a relatively robust (though still evolving) understanding of its target customer segment, with appropriate technology in place, and with high-powered incentives to motivate the creation of the content that would attract customers, DoCoMo would quickly be able to begin to benefit from network externalities, positive feedback, and dynamic increasing returns.

Significantly, in its own literature DoCoMo makes reference to the importance in its system of 'positive feedback' loops, although it is unclear whether this property was consciously 'designed into' IIS at an early stage or whether it has emerged as an *ex post facto* description of the innovation system that has evolved. Be this as it may, it is clear that network externalities, positive feedback, and dynamic increasing returns are a key determinant of the dynamic efficiency properties of IIS.

Network externalities arise when the value of participation in a network increases as more people join that network. In the case of i-mode, the more customers that sign up for the service the greater the ability of each customer to communicate with others (for example, by sending e-mail messages or pictures to other i-mode customers). Furthermore, the greater the number of customers, the greater the incentive for content creators to generate content and for complementary equipment makers (such as handset manufacturers) to produce equipment. It is here that the positive feedback loops kick in as the greater availability of innovative content and equipment for i-mode, the more new customers are likely to be attracted to the service, leading to further spirals in the virtuous circle. The end result of this virtuous process is dynamic increasing returns that accrue as improvements are made in technology and organization which imply over time falling unit costs and increasing quality. In turn this tends to give a competitive advantage to the firms lucky enough to benefit from this virtuous cycle. In this way DoCoMo was able to join other firms such as Microsoft and Intel (Wintel) and Matsushita (VHS) that have benefited from similar advantages.

In DoCoMo's case this explains the *increasing* lead that the company has been able to establish over its rivals in Japan. This lead may be measured in terms of number of subscribers, total revenue from i-mode services, or number of i-mode compatible web sites, including both sites appearing on the i-mode menu (i.e. i-menu sites) as well as sites that are simply i-mode compatible (see below).

Conclusion

It is these four properties of IIS *in combination* that explain the success of DoCoMo's i-mode service relative to both the company's other competitors in Japan and to counterparts in Europe and the US. However, it is one thing to identify the properties of a system that have enabled it to become an economically efficient system but quite another to explain how that system came to be. The following section, therefore, will provide a brief account of the emergence and evolution of i-mode and the IIS.

HOW DID THE 'i-MODE INNOVATION SYSTEM' EMERGE AND EVOLVE?

A brief summary of the i-mode evolutionary path is provided in Table 9.3. This is followed by a more detailed account of the emergence and evolution of i-mode and the i-mode innovation system.

The Starting Point: Mobile Voice Running Out of Steam

How did the i-mode service and IIS emerge and evolve? The starting point, as shown in Table 9.3, was the realization by Mr Ohboshi in 1995–96, then President of DoCoMo (which became semi-autonomous from its parent, NTT, in 1992), that something had to be done to counter the effects of longer-term declining growth in revenue from mobile voice communications. This decline in growth was the inevitable result of saturation of the mobile voice market as the diffusion of mobile services approached their limits and the uses of voice communications became fully satisfied. His general answer to this problem was to develop new mobile data services. (Significantly, the other Japanese mobile operators were simultaneously coming to the same conclusion as is evident from the fact that they introduced their own mobile Internet services only months after DoCoMo.)

Constituting the Development Team

In order to give concrete effect to the search for new mobile data services Mr Ohboshi asked Mr Kei-ichi Enoki to establish a team that would develop these services. Enoki's open-mindedness in carrying out his task is indicated by

Table 9.3. Periodizing the evolution of i-mode

Date	Event
1995–96	Mr Ohboshi (then President, now Chairman, of DoCoMo) decides that new data services are required to complement falling growth in mobile voice revenue. He requests Kei-ichi Enoki, a DoCoMo employee, to establish a team to design a new mobile data service.
1996	Mr Enoki begins work. He hires Takeshi Natsuno, a former Japanese Internet startup manager, and Mari Matsunaga, former editor-in-chief of a jobs magazine owned by Recruit. The team rejects the original idea of targeting the business market in favour of the young mass consumer market. Matsunaga proposes the 'magazine metaphor' to shape decisions about content and pricing for the new i-mode service while Natsuno draws on 'Internet thinking' to create incentives for content and application providers. Enoki complements this, drawing on his experience with his own children.
1997	DoCoMo's DoPa packet transmission service started.
1997–99	The team iteratively makes key decisions about matters such as evolving customer tastes and preferences, pricing, choice of appropriate technologies, contents required and incentives for content creation, and the development of enabling equipment such as handsets.
22 Feb. 1999	i-mode service started.
March 2001	20 millionth customer signed up.
	1,420 recognized i-mode sites available through i-mode menus.
	39,119 unrecognized i-mode-compatible sites.

Source: M. Fransman.

the fact that amongst those he hired was Mrs Mari Matsunaga, then editor-in-chief of a magazine that advertised jobs owned by a company called Recruit,[11] and Takeshi Natsuno, a former Japanese Internet start-up manager. To go outside a large, established Japanese company was by no means an obvious choice in Japan where labour markets in large companies are mainly internal. As things turned out, however, Matsunaga and Natsuno were to make a major contribution to the cognitive framework that was to play a key role in shaping the evolution of IIS.

Re-Targeting the Market and Pricing

In considering the key question of targeted market segment the team soon came to the conclusion that rather than a business market it was preferable to

---

[11] This section is based on the author's personal interview with Mrs Matsunaga. She has since written her own account of her i-mode experience: Matsunaga, M. (2000). I modo jiken (The i-mode Event).

aim for the young mass consumer market.[12] This conclusion was also influenced by Mr Enoki's observation of the way his own children were using pagers for the purposes of communication with their friends.

Having made this crucial switch in targeted market segment, Mrs Matsunaga's experience in the Japanese magazine world became an important reference point, referred to by the present author as the 'magazine metaphor'. Mrs Matsunaga suggested that the new mobile data service should provide the kind of content that was available in the magazines that young Japanese were already buying in large number.

The 'magazine metaphor' also had significant implications for the pricing of the new service. It accordingly came to be felt that the new service should not cost more than the average magazine and eventually a maximum monthly charge of 300 Yen (about $3) was agreed. In addition, as shown in more detail later, customers would be charged the normal price for their mobile voice service, as well as a bit-rate (rather than time rate) for the information they sent and received. Customers would also be able to purchase content from some of the web sites they could access via their i-mode menu (so-called i-menu sites). Consistent with the guidelines suggested by the 'magazine metaphor', DoCoMo decided that the maximum charge for i-menu contents should not exceed the monthly charge for the i-mode service as a whole. Accordingly, it was agreed that i-menu content providers would be given three charge options: 100, 200, or 300 Yen, all on a monthly basis.

High-Powered Incentives for Content Creation

DoCoMo would use its billing system to collect these micro payments and would deduct 9 per cent for the 'billing service' that it provided to the content providers. This meant that the providers were left with 91 per cent of the revenue generated from their content, thus creating what has been termed in this chapter a high-powered incentive to provide content. In addition, benefiting from the economies of scale provided by DoCoMo's billing system, content providers would be given a low-cost way of providing for transactions payments. Furthermore, they would benefit from low risk (a customer refusing to pay would likely lose all mobile communications facilities) and from the security given by DoCoMo's network.

With high-powered incentives in place, and with DoCoMo the market leader in cellular communications in Japan (in 1997 DoCoMo's market share was 53 per cent, rising to 58 per cent in 1999 when i-mode was introduced), there were important reasons motivating content providers to create content for

---

[12] Significantly, in early 2001 Dr Tachikawa, current CEO of DoCoMo, announced that the company would target the business market for its first third-generation mobile services, brand named FOMA, introduced at the end of 2001.

i-mode. For DoCoMo, IIS provided important additional benefits. By drawing a clear distinction between i-menu sites—that is sites that could be directly accessed by customers via the i-mode menu on their handsets—and other 'unrecognized' sites that were i-mode compatible and could be reached indirectly by inputting the site's Internet address, DoCoMo gained the ability to make its own selection of content-providing sites that it judged to be superior. I-menu sites, being on the i-mode menu, all other things equal, have a higher probability of being 'hit' by customers. DoCoMo's billing service facilitating micro-payments was only made available to the i-menu sites.

This procedure provided further dynamic benefits. To begin with, it meant that DoCoMo did not have to try and forecast customer tastes and preferences. Instead the company could run its own giant experiment, by initially selecting large numbers of i-menu sites and leaving the final-round selection to be made directly by customers 'voting' with a press of their i-menu buttons. In this environment consumer tastes and preferences would be emergent rather than forecast and DoCoMo, together with the successful content creators, would be able to appropriate the benefits. In addition IIS created a process of competition between content providers thus creating further pressures and incentives for effective content creation.

Choice of Appropriate Technologies

The i-mode team also had to make choices regarding appropriate technologies. One choice that had to be made related to the transmission technology that would be used. As noted in Table 9.3, here the team was assisted by DoCoMo's DoPa packet transmission service that began in 1997. Packet transmission (as Europe's first-generation WAP providers discovered to their considerable cost) provided important benefits from the customer's point of view. These benefits included, notably, the always-on functionality provided at the press of the handset's i-mode button as well as the ability to charge the customer by bit rather than by time, thus considerably reducing the overall cost of accessing web sites. Furthermore, packet switching facilitated the development of billing systems that could provide for micro-payments, a facility that only became available in 2.5-generation European mobile networks with the introduction of GPRS (general packet radio service) in 2001–02.

A further technology choice had to be made regarding the way in which contents on fixed Internet sites would be made available to i-mode users. Here too DoCoMo made a choice that was particularly user-friendly, although in this case the users concerned were content creators rather than final customers. By choosing the existing Internet standards based on HTML, HTTP, and TCP-IP, DoCoMo and its partners came up with cHTML (compact HTML) which was already largely familiar to Internet web site creators and software companies.

(cHTML was originally designed by a Japanese company established in Tokyo in 1984 called Access which collaborated with the Japanese electronics companies, Matsushita, NEC, Fujitsu, Mitsubishi Electric, and Sony. Access also developed the micro-browser that enables Internet access in all DoCoMo's i-mode phones.)

By basing its technology on knowledge that was already widely distributed, DoCoMo could take the next step and develop its own web site giving freely the information and tools needed for potential content creators to develop their own i-mode compatible sites and either bid to become an i-menu site, accessible directly from the i-mode menu on the user's handset, or remain 'unrecognized' but still accessible to i-mode users. As shown in Table 9.3, by March 2001, with 20 million i-mode customers signed up, there were 1,420 recognized i-menu sites and 39,119 unrecognized sites.[13] As these figures make clear, IIS has been outstandingly successful in lowering the barriers to entry facing new content creators thus providing a considerable boost to the dynamic effectiveness of the system as a whole.

Marketing i-Mode

In marketing i-mode the team took the important decision not to emphasize 'surfing the Internet'. It appears there were several considerations motivating this decision. The first was that the experience of Internet surfing was not nearly as widespread in Japan as in the US. Japan had a personal computer penetration rate of around 20 per cent compared to the US's 44 per cent. Furthermore, it was by no means clear that fixed Internet web pages would look good in their reduced form on a small mobile handset display. Similarly, it was also unclear whether 'Internet surfing' was commonly perceived as a satisfying consumer experience. Accordingly the team decided instead to focus on the fun and useful content, information, and applications that were accessible via i-mode. (Significantly, DoCoMo's European counterparts did not come to the same conclusion, choosing to emphasize the ability to 'surf the Internet' provided by WAP-enabled phones.)

In Table 9.4 information is provided on the popularity of different types of content accessed through i-mode.

New Modes of Consumption

Although the 'magazine metaphor' played an important role in the cognitive framework that shaped the development of i-mode, the metaphor also has its

---

[13] Information for potential content creators, referred to by DoCoMo as IPs (Information Providers), is available on http://www.nttdocomo.com/i/tag/newip.html.

Table **9.4**. Consumption of i-mode applications by type, 2001

Application	Percentage of users
Entertainment	40
News & Weather	15
Tickets & Living	11
Share Trade	8
Travel	8
Dictionary & Phone Book	7
Other	9

limitations. The most obvious limitation is that while magazines belong to the world of print, i-mode is an electronic medium that operates under a specific set of production and consumption environmental conditions. The case of 'niche-time' content development makes this important distinction clear. One example of such development is the production of games that last only a minute or two and are designed to fit into short time niches such as those that occur when a person is commuting. Another example is the cartoon character Itsudemo Charappa created by the entertainment company Bandai which for a mere 100 Yen would appear each morning on the user's i-mode handset, netting Bandai more than $1 million per month. Clearly, this kind of capability is not available in the print world.

Examples of i-Mode Content Sites

Some examples of i-mode contents sites are given in Table 9.5.

The Cognitive Contribution

It is clear that the cognitive framework that was adopted by the i-mode team played a crucial role in shaping the evolutionary outcome. The cognitive framework that was adopted, however, was itself determined by several key influences. To begin with, as already discussed, the context within which the cognitive framework was constructed was determined by the fact that NTT DoCoMo was a mobile network operator searching for new revenue streams to complement those from mobile voice. This determined the overall objective which was to develop new data-based Internet services.

Within this context, however, the personal cognitive frameworks of the leading members of the development team had an important impact both on the processes that led to the ultimate i-mode concept as well as the final

Table 9.5. Some i-mode content sites

Contents	Site
Make a homepage for free	Magic Island
Find a friend	i-ChaT Club
Keep up to date on friends' activities	maHima
Download ring melodies and set up ID ringing	J's Communication
Find a love hotel	Boutique Hotel Search
Locate a hostess bar (18 + years only)	Big Desire
Do a sake search	Japanese Sake Data Bank
Get a wake-up call	Friend's Call
Have your fortune told	J-Skyweb Fortune Teller
Research shares	i-Stock
Check national train timetables	JR Travel Navigator
Check traffic	ATIS Traffic Information
Make hotel reservations	i-Hornet
Make flight reservations	ANA Reservations
Rent a car	Nippon Rent-a-Car
Find out if you're crazy	Psychological Evaluator
Calculate taxi fares	Taxi Fare Database
Find emergency numbers	i-Mode Useful Diary
Look up a Japanese word	Online Dictionary
Find a restaurant anywhere in Japan	Restaurant Finder
Get marriage counselling	@Heart Clinic
Master Pachinko	GIC 'How To Win' Info
Find your soul mate	Cell Phone Match-Maker

Source: J@pan.Inc, June 2000.

outcome. For example, Natsuno has referred to what he has called the 'telecom way of thinking'[14] which he argues is prevalent in all telecoms companies, including DoCoMo. This he contrasts with his own 'Internet way of thinking' which formed a crucial part of his experience and learning when he was employed by a Japanese Internet start-up company prior to joining DoCoMo. He argues that it was this Internet way of thinking that led him to understand (a) the importance of involving independent content providers in the project, (b) the necessity to give them strong incentives so as the create a win–win situation for both DoCoMo and the content providers, and (c) the desirability of having low technology barrier to entry for these content providers. The latter determined the i-mode team's choice of compact HTML as the language for i-mode web sites, a version of a language that was already widely known and understood since it was a standard language used in the Internet. In Natsuno's words: 'With the Internet the first and shortest path to maximizing profit is to maximize third-party content providers' profit. So always I emphasize that business models should be flexible enough to provide an incentive to others,

[14] Interview with Takeshi Natsuno quoted in J@pan.Inc, June 2001, p. 36.

and if so, it's going to lead finally to our business profit. That is what I proved with i-mode.'

Furthermore, the incentive, to be effective, needed to be not only financial but also required low technological costs of entry: 'Only by providing a real business model, a real incentive for them [i.e. the content providers], and by lowering technology entry barriers for them to join this platform—by using de facto standard technology—was I able to get 67 companies to provide content for i-mode's first day.'[15]

Similarly, the contribution made by Mrs Matsunaga's 'magazine metaphor', a central part of her cognitive framework, also played an important role in shaping the processes and outcome of the i-mode team's work as noted earlier.

THE VERY DIFFERENT EUROPEAN EVOLUTIONARY PATH[16]

In Europe, as noted in the introduction to this chapter, a very different evolutionary path towards the mobile Internet was followed. This evolutionary path is traced in this subsection.

To begin with, in Europe the initiative was taken, not by mobile network operators as in Japan, but by the dominant specialist mobile equipment suppliers. Ericsson and Nokia—after making their own internal efforts to create ways of reformatting web pages to facilitate Internet access from mobile phones—eventually teamed up with other companies. These included Unwired Planet and Motorola from the US.

Second, the cognitive framework governing the exercise was fundamentally different in Europe from that chosen in Japan. In Europe, after the equipment companies decided to join forces, the objective was to develop a global de facto *standard* for mobile Internet access. However, in Japan the objective was to develop a new mobile-based *consumer service*. It is for this reason that DoCoMo's Natsuno, looking back, concluded that 'i-mode is a business model, not a [technological] system' like WAP.[17]

It was the construction of these two very different cognitive frameworks, more than anything else, which gave rise to the divergence of the two subsequent evolutionary paths. In turn, the choice of these cognitive frameworks was a function of the different role in the Telecoms Industry and the interests of the companies that created the frameworks.

The role of the mobile equipment supplying companies, as shown in the Layer Model developed in Chapters 1 and 2, was to provide technologies embodied in

---

[15] Interview with Takeshi Natsuno, op. cit., p. 36 and p. 37 respectively.

[16] This section draws heavily on the excellent paper by Jon Sigurdson, which provides the first detailed account of the development of WAP (wireless application protocol). The paper, 'WAP OFF—Origin, Failure and Future', can be downloaded from http://www. TelecomVisions.com.

[17] Excerpt from Takeshi Natsuno's book, *i-mode Strategy*, Nikkei Business Press, 2000, quoted in J@pan.Inc, June 2001, p. 36.

equipment to mobile network operators. Their role was *not* to develop new mobile services for final customers. That role, rather, was played by the mobile operators. In short, the customers of the mobile equipment customers were the mobile operators. The customers of the latter were final users.

In Japan, however, NTT DoCoMo was a mobile operator. As such, as already noted, it was searching for a new stream of revenue that would complement the slowing revenue and profit growth from mobile voice services. The same was true of DoCoMo's competitors, now KDDI and J-Phone. Their interests, therefore, lay in the development of new mobile services.

There were other important determinants of the Japanese evolutionary path. The first was that DoCoMo was in a stronger position than its European counterparts to influence the choice of technologies that would be used to deliver mobile Internet services, including both network and handset technologies. As part of the NTT Group, DoCoMo came from an extremely strong technological tradition. The reason is that NTT had always, from its beginnings in the nineteenth century, had strong in-house R&D competencies and worked closely with a small group of equipment suppliers. This put NTT in an extremely good position to shape the research and development processes that determined the kind of equipment and handsets that were produced. DoCoMo embodied this same tradition. This can be seen clearly from its R&D intensity which is far higher than that of its counterparts such as Vodafone.[18]

In Europe, on the other hand, as a result of the process of division of labour and vertical specialization examined in Chapters 1 and 2, the mobile network operators tended to be far more dependent on the initiatives of the specialist mobile equipment suppliers. This is evident in the relatively small degree of influence the operators have had in shaping the design and development of mobile handsets[19] and new mobile technologies such as GPRS and EDGE.[20] It is no exaggeration to say that in Europe it is the mobile equipment companies that have been proactive in developing the new mobile technologies, while the mobile operators have tended to be more passive, even though as large customers

---

[18] For a very detailed analysis of the evolution of NTT from the late nineteenth century, and the reasons for its particularly strong technological competencies, see Fransman, M. *Japan's Computer and Communications Industry* (Oxford University Press, 1995). In 2000 DoCoMo's (as opposed to NTT's) R&D intensity (R&D as a proportion of sales) was around 4.5 per cent. Vodafone, however, was not included in the *Financial Times* R&D Scoreboard, 27 September 2001. The reason is that Vodafone has tended to outsource much of its R&D requirements, particularly to Ericsson.

[19] One notable consequence is that Japan has had a significant lead over Europe in developing mobile handsets with colour screens and with functionalities such as in-built digital cameras. This, however, may also be a function of the greater degree of competition in Japan amongst handset developers than exists in Europe or the US. (Until now, Japan has operated under a different second-generation mobile standard than Europe or the US and this has meant that Japanese handset companies have not been able to sell their products, developed for the Japanese market, in Europe and the US. This will change, however, with the acceptance of W-CDMA as the standard in Europe, Japan, and parts of the US. See footnote 5 above for an explanation of W-CDMA.)

[20] See footnote 5 above for further information on GPRS and EDGE.

and as important users they have had an influence over the process. In Japan, although the same tendency exists, it is not nearly as pronounced as in Europe. DoCoMo's important role in shaping the technologies used for i-mode, including handsets, provides strong evidence in support of this contention.

But this explanation of the divergent evolutionary paths in Japan and Europe leaves one crucial question unanswered. Why did non-Japanese mobile network operators—such as Vodafone, Deutsche Telekom's T-Mobile, France Telecom's Orange, Telecom Italia's TIM, as well as the US mobile operators—not confront the same questions at the same time as DoCoMo's Mr Ohboshi and why did they not come to the same conclusion? Surely they too could see from the latter 1990s that time was running out for mobile voice services as the main engine for growth in the mobile industry. Did they not also realize from this time that steps would have to be taken to develop mobile data services of various kinds and that mobile Internet was one important way forward that had to be explored? In short, why did they not make their own efforts, with or without their mobile equipment suppliers, to develop new mobile Internet *consumer services* such as i-mode?

This important question poses a further puzzle since there seems to be little evidence of the non-Japanese mobile operators being highly proactive in places such as the WAP Forum in attempting to develop such services. Further research is required to answer this puzzle convincingly. However, there is an hypothesis that might help to solve the puzzle.

This hypothesis draws on the analysis of vertical specialization presented in Chapters 1 and 2 of this book in the discussion of the Layer Model. It is based on the relatively weak R&D competencies of the mobile network operators and the corresponding comparative R&D strengths of the equipment suppliers.

This hypothesis argues that precisely because the non-Japanese mobile network operators had tended to leave the R&D function to the specialist mobile equipment suppliers, and because the development of new mobile Internet consumer services required significant R&D expenditure and capabilities, the operators preferred to wait until the equipment suppliers came up with a solution. When WAP eventually arrived, the operators tried to sell WAP-based services. However, since WAP was developed as a *standard* rather than a *service*, its functional characteristics from the consumers' point of view were inappropriate and, equally important, there was insufficient WAP-based content and applications to give users an incentive. The resulting failure of WAP, given the different cognitive framework that shaped its development, was therefore inevitable. This hypothesis needs to be further tested.

MAKING MONEY FROM i-MODE

How much money does DoCoMo make from i-mode? Specifically, what are the revenue streams that are generated by i-mode and how much is derived from each of these streams? These important questions are answered in Table 9.6.

Table 9.6. DoCoMo's revenues from i-mode

Revenue	Monthly amount (Yen)
i-mode revenue streams	
Monthly i-mode fee per user (all DoCoMo's phones now include i-mode)	300
Packet charge (fee to send/receive the 128 byte packets, Yen 0.3 per packet), average per month per subscriber	1,000
Premium service (e.g. financial information, or character download developed by Bandai), average per month per subscriber	100
Additional voice, average per month per subscriber (i.e. additional voice as a result of i-mode that would not have taken place otherwise, e.g. phoning restaurant to make booking after locating restaurant on i-menu)	1,500
Total	2,900
DoCoMo's average revenue per user (ARPU) for traditional voice only, March 2000[1]	8,620
i-mode revenue stream divided by traditional voice only ARPU	34%

As noted in the text, premium services are charged at the rate of either 100, or 200, or 300 Yen per month of which DoCoMo receives 9 per cent.
[1] ARPU data from http://www.nttdocomo.com/ir/operate.html.
Source: M. Fransman.

Table 9.6 shows two things: the sources of revenue generated for DoCoMo from i-mode and the importance of this revenue relative to the average revenue per user (ARPU) for the company's traditional cellular voice service.

i-mode Revenue Streams

As Table 9.6 shows, i-mode generates four revenue streams for DoCoMo. The first is from the monthly i-mode fee of 300 Yen (about $3) paid by subscribers. All DoCoMo handsets (manufactured by a number of Japanese electronics companies but sold under the DoCoMo brand name) come with the i-mode button that provides immediate, always-on access to the i-menu sites. The second revenue stream comes from the packet charge that DoCoMo levies on its subscribers. They are charged 0.3 Yen per 128-byte packet, yielding an average of 1,000 Yen per subscriber per month.

The third revenue stream comes from DoCoMo's commission on revenue generated by i-menu sites, that is the recognized sites that are directly accessible from the i-mode menu. The company's commission, charged in return for the billing and micro-payment service that it provides to its recognized content providers, amounts to 9 per cent of the total revenue received by the provider. On average this yields about 100 Yen per subscriber per month. The fourth and final revenue stream comes from the *additional* voice revenues that result from the use of i-mode services. An example is a phone call to book a restaurant that has been located on the i-menu. On average, additional voice revenue amounts to 1,500 Yen per subscriber per month.

The bottom part of Table 9.6 shows that the average revenue stream per user per month from i-mode services amounts to 34 per cent of the average monthly revenue per user for DoCoMo's traditional cellular voice services.

DoCoMo's Benefit from i-Mode

Several important conclusions emerge from Table 9.6. The most important is that the bulk of DoCoMo's revenue from i-mode comes, not directly from the content that is provided, but from the network traffic that is generated by customers accessing this content. More specifically, 52 per cent of the total i-mode revenue stream (1,500 out of 2,900 Yen) comes from additional voice traffic, while 35 per cent (1,000 out of 2,900 Yen) comes from charges on the packets of information that are carried as a result of the i-mode service. This means that 85 per cent of DoCoMo's revenue from i-mode comes from the network traffic generated. Only 10 per cent comes from the flat-rate i-mode monthly fee (300 out of 2,900 Yen) and only 4 per cent comes directly from the content/billing micro-payment service that DoCoMo provides to i-menu content providers.

However, it is DoCoMo's publicly declared intention to build on this first business model and to extend its sources of revenue. One way in which this will be done is through the generation of advertising revenue. In order to further this strategy in June 2000 DoCoMo established a company called D2 Communications together with the Japanese advertising company, Dentsu, and NTT Advertising. The objective of D2 Communications is to generate advertising revenues from the i-mode platform. A further source of revenue comes from extending DoCoMo's billing and micro-payment service to other methods for facilitating e-commerce transactions. To take this further in March 2000 DoCoMo invested in an Internet-based online bank, The Japan Net Bank, one of the main shareholders of which is the Sakura Bank. In addition, in June 2000 DoCoMo invested in Payment First, an internet payment services company mainly owned by the Japanese electronics company, Oki Electric.

As the present discussion makes clear, therefore, DoCoMo's i-mode service can be understood, not simply as a mobile Internet service, but rather as a *portal platform* that is being used to launch a wide range of content, applications, and services.

Table 9.7. DoCoMo's global strategy

Strategy
Establish common W-CDMA platform with global partners. Capture global growth potential by actively promoting mobile internet services. Increase DoCoMo's and partners' competitiveness through content sharing and co-sourcing.

Source: DoCoMo.

DoCoMo's GLOBALIZATION[21]

Reaping the benefits of network externalities, positive feedback loops, and dynamic increasing returns in its home Japanese market, DoCoMo by 2000 had set the stage for its future globalization. DoCoMo's global strategy is depicted in Table 9.7.

Several comments may be made regarding DoCoMo's global strategy as outlined in Table 9.7. The first is the importance the company attaches to the establishment of W-CDMA as the global standard for third-generation mobile services, a standard to which DoCoMo has made significant contributions and in which it is a leader. For this strategic objective to be achieved, other rival standards will have to be prevented from becoming globally dominant. The latter include CDMA 2000, a standard championed by the US company Qualcom which is being adopted by several mobile operators in the US and Korea. As will be seen later, a key motive behind DoCoMo's acquisition of a 16 per cent holding in AT&T Wireless was for the latter to adopt W-CDMA. Similarly, DoCoMo's other global partners in which the company has invested have also adopted W-CDMA.

Second, DoCoMo sees mobile Internet services as forming a key part, perhaps even the core, of third-generation mobile services. This, however, should not deflect attention from other services that will also be a significant part of the third-generation offering. In its own public literature, for example, DoCoMo also emphasizes the importance of person-to-machine communications as well as machine-to-machine communications (for example vending machines communicating to company computers that they need to be replenished).

---

[21] Since writing this chapter, DoCoMo has taken a significant financial knock as a result of having to write down the value of many of its overseas acquisitions. These acquisitions were bought at relatively high values as a result of the Telecoms Boom analysed in Chapter 1, only to see them rapidly depreciate during the subsequent Bust. However, this fate has also had a significant negative impact on all the main mobile network operators and equipment suppliers and at the time of writing DoCoMo is pressing on with its globalization strategy outlined here.

Third, based on its very positive experience with i-mode, DoCoMo understandably stresses the importance of content generation and sharing as a key part of its global strategy. In addition, it emphasizes the benefits that it and its global partners might receive from co-sourcing from equipment and technology suppliers.

A key element in DoCoMo's global strategy is the creation of a global platform that will play the same role internationally as its i-mode platform has played in Japan. Table 9.8 shows the main elements in this platform.

In order to implement its global strategy and establish its global platform DoCoMo has created a global network of strategic partnerships. These partnerships are shown in Table 9.9.

A striking feature of Table 9.9 is that DoCoMo has taken only minority holdings in its overseas strategic partners. This is a strategy that Vodafone also followed from 1988, the year the company began investing in foreign mobile operators, to 1998 when it acquired majority stakes in AirTouch and Mannesmann. Other telecoms companies, such as Deutsche Telekom, have

Table 9.8. DoCoMo's global platform

Platform module	Specifics
Network	GSM/GPRS (2G/2.5G)
	W-CDMA (3G)
Contents	Local/universal
	Text/non-text
Service platform	Dual browser (HTML subset + WML)
	Next generation WAP browser
User terminal	JAVA
	New devices (car navigation,
	PlayStation, etc)

Source: DoCoMo.

Table 9.9. DoCoMo's global partnerships

Partner	Country	DoCoMo's share (%)
Hutchison Telecom	Hong Kong	19
Hutchison 3G	UK	20
KPN Mobile	Netherlands + other European countries	15
AT&T Wireless	US	16
AOL Japan	Japan	42.3
KG Telecom	Taiwan	20
Telefonica Cellular—Tele Sudeste Celular	Brazil	4

Source: DoCoMo.

insisted only on majority holdings. DoCoMo, however, has argued that it will be able to secure a significant degree of influence over its partners' decisions, not through its shareholding, but rather through the contribution that it makes to the partner as a result of its technological expertise and business models. Minority holdings also allow DoCoMo to spread its investments and risks more widely globally, to buy time in learning how effective its partnerships are, and in economizing on its still growing stock of internationally experienced person-power. Less burdened by high prices for auctioned third-generation mobile licences than many of its European competitors—Japan chose the beauty-contest rather than the auction road to 3G—from 2001 DoCoMo began its active involvement in European mobile markets.

CONCLUSION

In this chapter a puzzle has been addressed. Why have attempts to introduce the mobile Internet been so much more successful in Japan than in Europe?

In tackling this puzzle it is shown that Japan's main success story—NTT DoCoMo's i-mode mobile Internet service—was so successful, not only because it offered superior functional characteristics, but also because it was based on an entirely novel innovation system. Like the Japanese just-in-time manu-facturing system, this 'i-mode innovation system' evolved in piecemeal fashion, its dynamic properties initially unintended and unforeseen by its creators. These dynamic properties are analysed in detail in this chapter.

However, it is one thing to analyse the properties of a system but quite another to explain how that system came to be. In the next part of the chapter an explanation is provided of how the i-mode innovation system emerged and evolved.

It is then shown that an entirely different evolutionary path towards the mobile Internet emerged in Europe. The divergence between the Japanese and European paths is explained in terms of the different cognitive frameworks that were created in each place. While in Japan the framework was shaped by the objective to create new mobile Internet *services*, in Europe the objective was to create a new global de facto *standard*. These different cognitive frameworks, in turn, reflect the fact that in Japan mobile Internet was created by mobile *operators*, while in Europe it was created by mobile *equipment suppliers*. A hypothesis is offered to explain why these different categories of player were in the driving seat in the two regions.

Cognitive frameworks also enter more directly into the i-mode success story. This is shown through the discussion of the contributions made by Takeshi Natsuno and Mari Matsunaga to the development work of the team that developed i-mode. Significantly, Kei-ichi Enoki, who was charged with the task of constituting the development team, went outside not only his company, but also the Telecoms Industry, in choosing two of the team's key members. Natsuno's background and learning was in a Japanese Internet start-up while Matsunaga's was in a marketing oriented recruitment company. Their contributions to the cognitive framework that was to shape thinking about the emerging i-mode innovation system are examined in detail. There is a strong parallel between this cognitive process and that examined in Chapter 2 involving the development of the key Internet-related technology, packet switching, originally opposed by telecoms engineers and researchers with a very different mindset.

Cognitive frameworks, of course, continue to be central in shaping the continuing evolution of the mobile industry. A key example is the major current issue regarding the kinds of next-generation mobile services consumers will want and how much they will be prepared to pay for them. The expected revenues from these services constituted an important assumption in the amount that mobile operators were willing to pay for third-generation mobile licences, an issue that has since become highly contentious. Significant interpretive ambiguity surrounds these issues.

Next an examination is undertaken of the revenue streams that have been generated for DoCoMo from its i-mode portal platform. It is shown that the bulk of the company's revenue has come, not directly from the content that has been provided, but from the extra traffic, both voice and data, that the i-mode service has generated over DoCoMo's mobile network.

Finally, it is shown how dynamic increasing returns, generated in Japan by the i-mode innovation system, have created the conditions for DoCoMo's globalization. The company's globalization strategies are analysed and the global partnerships that it has established examined.

In conclusion, this paper establishes that DoCoMo's i-mode is far more than only a successful mobile Internet service. In reality it is also a dynamic innovation system that endogenously generates over time significant co-evolving innovations in content, customer tastes and preferences, applications, enabling technologies and forms of organization. As such it is appropriate to include DoCoMo's i-mode system with other radical systemic innovations, such as the just-in-time manufacturing system.

Conclusion: Shareout, Consolidation, and the Restoration of Profitability

The Telecoms Boom collapsed under the weight of its own contradictions. These contradictions were analysed in Chapter 1 as flaws in the Consensual Vision that underpinned the Telecoms Boom. The glue that kept the Telecoms Boom going until 2000, despite the faults in the beliefs embodied in the Consensual Vision, was the overly optimistic expectations held by exuberant financial markets. Guided by these expectations, financial markets continued to pour finance into the Telecoms Industry on the assumption that future earnings would provide justifying returns. This raised telecoms share prices and created the self-fulfilling feedback loops that fuelled further increases. But, when the generalized financial boom ran out of steam after March 2000 and the faults in the Consensual Vision started to become apparent, 'market sentiment' changed making it harder to find the future earnings that had been expected. The market bull had turned into a bear.

The share prices of all telecoms companies, including the strongest such as the single-play mobile operator, Vodafone, collapsed. For many of the new entrants bankruptcy followed. Companies such as Global Crossing, Global Telesystems, Viatel, Winstar, Teligent, and Atlantic—that had been among the leaders in terms of the growth in share value—filed for protection. Some of these operators were bought by other stronger telecoms companies.

For example, KPNQwest, the joint venture between the Dutch incumbent and the US new entrant, bought Global Telesystems; Level 3 purchased assets from Viatel; and several prominent Asian telecoms companies are, at the time of writing, negotiating to acquire assets of Global Crossing. However, this process of the strong buying the weak, a normal process in any industrial shakeout, is constrained by the weakness of the strong caused by the dissipation of their core business, the undermining of their market value, and in many cases their continuing high levels of indebtedness (analysed in detail in this book). The process is also constrained by accumulating evidence that mergers and acquisitions often fail to deliver their intended goals. In turn, this has opened the door to private equity companies and others seeking to make fortunes from the knock-down asset prices of the unfortunate.

But for some the black clouds that have gathered over the Telecoms Industry have a silver lining (or at least can be construed as such in the hope of generating some new enthusiasm for telecoms companies). One example is COLT's current CEO, Peter Manning, safe (at least for the time being) as a result of the injection into the company of £494 million through a share issue in December 2001 backed by its founder, Fidelity, one of the largest US mutual funds. According to Manning the shakeout is good news for COLT 'in the mid- to long-term' because 'it obviously reflects the fact that there is a reduced competitive environment'.[1] Oligopolistic pricing would certainly give some relief to those companies that remain in the market, whether incumbents or new entrants.

However, many important questions remain. Will the network layer of the Telecoms Industry be dominated by a handful of global operators drawing their competitiveness primarily from economies of scale and scope? Or will continuing changes in technology together with regional variations in demand provide the basis for a fairly large number of companies remaining in the industry and specializing by function or region or both? What will the necessary conditions be—in terms of scale, scope, and structure—for those remaining in the industry? How large (with respect to what dimensions of largeness) will firms have to be to survive? Will they be able to offer the full range of services and be involved in several of the layers of the industry, or will competition force them to specialize more narrowly? Will the incumbents be able to remain as loosely vertically integrated companies, or will market pressures force them to break up? Will the activities of network operation and service provision become separated in different companies? There remains considerable interpretive ambiguity regarding questions such as these.

INCUMBENTS VERSUS NEW ENTRANTS

What will happen to the grand old incumbent network operators and their new entrant competitors whose story has been told in this book? As we have seen, at one stage the incumbents looked vulnerable to their focused, flat, fast, and flexible rivals. However, as the economic and financial realities of the late 1990s and the early years of the new millennium unfolded so it became clear that the traditional core businesses of the incumbents, though shrinking relatively, gave them a financial stability that many of the new entrants—facing extremely high fixed costs without sufficient immediately compensating revenues—lacked. Furthermore, the incumbents were able to take advantage of the new growth areas in mobile, the Internet, and data communications in order to try and

---

[1] *Financial Times*, 23 January 2002.

stabilize their positions. In short, by the beginning of the twenty-first century the old empire was able to strike back, an empire the origin of whose firms could be traced back to the late nineteenth century. Most of the major new entrants, whose futures had been so extolled by the financial markets during the Telecoms Boom, had been humbled by 2002 (as shown in Chapter 7).

And yet, as Chapter 8 on City of London Telecommunications (COLT) suggests, it would be unwise to dismiss all the new entrants. The best of them in time may be able to use their focused specialization, their organizational characteristics, and their strong positioning in key markets (particularly those prone to bottleneck) in order to grow and prosper in an environment of steady growth in demand for telecoms services. Furthermore, as just noted, they are likely to benefit along with the incumbents as consolidation and shakeout reduce the number of competitors along with the intensity of competition. In short, there are grounds for believing that the drama of the incumbent operators versus new entrants has not yet reached the final act.

GLOBALIZATION?

What about globalization? As the liberalization era was launched in the mid-1980s there were many in the industry who believed that a rapid process of globalization would be unleashed leading to a domination of the global Telecoms Industry by a handful of network operators and suppliers. However, by the early twenty-first century there was little sign, as this book shows in detail, of truly global network operators (with the significant partial exception of Vodafone[2]). The proportion of the revenue of operators coming from outside their home country or region remains small and the Telecoms Bust, if anything, has reversed the process of globalization by the main operators. Successful but relatively small regional operators—like Telia in Sweden or Sonera in Finland—though facing their share of problems, seem to have managed to survive and have not yet been swallowed by their larger and more powerful neighbours such as Deutsche Telekom and France Telecom. All the main European incumbents have significantly scaled back their global ambitions and, for the moment at least, have reconciled themselves to being regional players at best.

But does this mean that a significant global concentration, such as has occurred for example in the global motor car industry, will not happen amongst operators in the Telecoms Industry? Or will the same tendencies emerge in this industry? The answers to these questions are not yet clear.

More clear, however, is that the situation regarding globalization is significantly different in the equipment layer of the industry. Here a few companies such as Cisco and Nokia have managed to grab significant global market share, even though their continued dominance remains, in the longer term at

---

[2] For a detailed study of Vodafone, see M. Fransman, *Why Vodafone?*, 2002 (mimeo).

any rate, contestable. In this part of the Telecoms Industry the parallels with other globalized industries such as motor cars seems far closer.

THE FUTURE OF THE LARGE CORPORATION

Does the Telecoms Industry have anything to teach regarding the future evolution of the modern large corporation?[3] Telecoms network operators have featured prominently in the list of the largest global companies. How have they changed over the 1990s and into the new millennium as they have adapted to their new circumstances?

Several tendencies emerge clearly in the present book. To begin with, as was shown in Chapters 3, 5, and 6 the incumbent network operators—including the Big Five, AT&T, BT, Deutsche Telekom, France Telecom, and NTT—have segmented their activities into relatively autonomous businesses coordinated by a head office under the control of the CEO and Board. Over the period, however, the head office was substantially reduced in size. Although different principles of segmentation were often applied, with the result that the markets addressed and the technologies used by the different businesses were not always consistent amongst the incumbents, there were some commonalities.

The most obvious commonality was that all of the incumbents have established distinct mobile businesses. In the case of BT, this mobile business (originally called BT Cellnet, later renamed mmO2) was separated completely from the rest of BT and floated. At the time of writing, AT&T intends to do the same with AT&T Wireless. NTT's mobile subsidiary, NTT DoCoMo, was organizationally separated from the rest of the company in 1992 (largely as a result of pressure from the Japanese regulator). However, DoCoMo remains, with NTT's four other major subsidiaries, under the control of NTT's holding company. Deutsche Telekom and France Telecom (FT) have also established their mobile subsidiaries (T-Mobile and Orange, respectively) as relatively autonomous businesses but have kept them loosely integrated under the control of the company's CEO and Board. As detailed in this book, all the Big Five incumbents underwent significant corporate reorganizations in the late 1990s and early years of the new millennium.

This raises important questions. Is it still correct to conceive of these companies as single large corporations? Or is it more accurate to see them as several companies (specializing in areas such as mobile communications, Internet-related services, business services, consumer services, etc), even though these companies are coordinated by a single overall CEO and Board? Certainly, NTT's DoCoMo, FT's Orange, and Deutsche Telekom's T-Mobile—relatively

---

[3] For recent popular discussion see, for example, 'The Future of the Company', *The Economist*, 22 December 2001, pp. 82–84, and 'Will the Corporation Survive?', *The Economist*, 2 November 2001, pp. 16–20.

autonomous subsidiaries after their company reorganizations—look far more like Vodafone than they did before. And their Internet subsidiaries look far more like AOL-Time Warner than they did before.

But there remains a crucial difference: the heads of these mobile subsidiary businesses still need to clear major decisions regarding strategic direction and large investments with the overall CEO and relevant corporate executive committees, unlike the completely autonomous companies such as Vodafone and AOL-Time Warner that do not have an additional organizational layer above them. Therefore, though the main subsidiaries of the incumbent network operators have come to be far more like their more focused single-play competitors, they are not yet identical to them.

Will there be a convergence? Will the subsidiaries of the incumbents in time be completely separated? My interviews with CEOs, board members, and senior executives of many of the main incumbent network operators suggest that (with the possible exception of AT&T and BT) the general feeling among the incumbents is that the positive interdependencies and synergies between the various businesses (e.g. between the mobile, Internet, and data businesses) are sufficiently great to more than compensate for the added organizational complexity that integration—however loose—implies. To the extent that this sentiment continues, the incumbents will remain as highly segmented, but still to some extent integrated, corporate entities.

THE IMPORTANCE OF KNOWLEDGE

The analysis in this book of the dynamics of change in the Telecoms Industry in the Internet Age has also revealed the fundamental role of knowledge. More specifically, it has revealed the importance of the way in which knowledge is created and used. In the pre-liberalization era, as was shown, a good deal of the knowledge required in telecoms was created in the central research laboratories of the main incumbent operators (i.e. knowledge relating to switches, transmissions systems, customer premises equipment, etc).[4] This knowledge was passed on to the specialist telecoms equipment suppliers, working closely with the incumbent network operator, who used it to manufacture network equipment. The equipment was then configured by the operator into telecoms networks used, in turn, to provide a range of telecoms services.

Over time, however, as shown in Chapter 2, two important changes occurred in this system for creating and using knowledge. First, the innovative competencies of the specialist equipment suppliers increased both absolutely and relative to those of the incumbent network operators. This was the result of

---

[4] A detailed study of the knowledge-creation processes in telecoms switches and optical fibre systems is to be found in Fransman, M., 1995, *Japan's Computer and Communications Industry*.

learning processes as well as the competition these equipment suppliers faced, both in their home market (though here they often enjoyed privileged access to the markets of the incumbent operator) and in third countries including developing countries. The best indicator of this change is to be found in the increasing R&D intensity of the equipment suppliers and the corresponding decreasing intensity of the incumbent operators (detailed in Chapters 1, 2, and 9).

Second, with the advent of the Internet, a powerful new platform for innovation was created that opened up the knowledge-creation process to many new kinds of innovators. The advantageous characteristics of the Internet innovation platform, together with the incentives that it provided, greatly accelerated and widened the knowledge-creation process.[5] Accessibility to the knowledge-creation process was greatly increased. No longer was this process monopolized by incumbent network operators and their equipment suppliers. (These changes are analysed in Chapter 2.)

These changes in the knowledge-creating system underlying the Telecoms Industry radically transformed the ways in which the created knowledge could be used. Whereas in the pre-liberalization era the user of the knowledge (the monopolist network operator) was also its creator (through the R&D undertaken in its central research laboratories), now new users of knowledge could emerge[6] (the new entrant network operators) without having to create the knowledge necessary (through in-house R&D).

The fact that specialist equipment suppliers made the latest knowledge (embodied in their equipment) easily available to whoever could pay for it revolutionized the dynamics of the Telecoms Industry. Furthermore, the fact that this knowledge was 'black-boxed' (so that the user of the knowledge needed no knowledge of what went on inside the box), and was modularized, which meant that only knowledge of the interfaces between the modules was needed, greatly facilitated the task of the engineers of the new network operators whose business it was to configure and operate telecoms networks in order to provide their services.[7] In short, the knowledge needed to use knowledge in the new Telecoms Industry was significantly less than that which was required in the old (pre-liberalization) industry. The competencies needed to use knowledge were reduced and this decreased the cost of access to knowledge.

As was shown in many parts of this book (most explicitly in Chapters 1 and 2) the effect of these changes in the conditions of knowledge-creation and use was to significantly reduce the technological entry barriers into the network operator layer of the Telecoms Industry. In turn, this increased entry and therefore the

---

[5] For the example of live video over the Internet, see Fransman (2000a), pp. 33–34.

[6] Helped by the new regulations that attempted to facilitate new entry and competition.

[7] In Chapters 2 and 7 figures were given on the number of network engineers employed by new entrants Energis and City of London Telecommunications (COLT), respectively, pointing to the relatively small number of network engineers required by the new entrant network operators.

intensity of competition in the telecoms services markets involved.[8] The entry of many vigorous new network operators had further important repercussions, as shown throughout this book, including during the Telecoms Boom and subsequent Bust.[9]

This book has also shown that the 'issue of knowledge' was not confined to the creation and use of knowledge involved in the provision of telecoms equipment and services. Another kind of knowledge issue emerged in attempting to know how the Telecoms Industry worked, how it was changing, and what a firm needed to know and do in order to survive and prosper in the industrial environment. These questions faced not only network operators and telecoms equipment suppliers but also, as shown in Chapters 1 and 2, financial analysts and investors in search of attractive financial returns from the industry. Here the knowledge-creating problem was how to understand the industry as a whole and its workings as opposed to creating the knowledge needed to make and competitively sell its products and processes. The 'solution' to this problem, as we have been at pains to stress throughout this book, was to construct cognitive visions embodying the key assumptions and beliefs of the creator. However, the problem with this 'solution' was that it often did not provide a solution. Mistakes required 'vision revision' by modifying and adapting the assumptions and beliefs in the cognitive vision. These kinds of knowledge processes, it has been shown throughout this book, were an inherent part of the process of evolutionary change in the Telecoms Industry in the Internet Age.[10]

Many questions have been raised in this concluding section, and doubtless there are many more that should have been raised. However, in a book premised on the assumption of uncertainty, ignorance, and mistakes it is perhaps unsurprising that no clear-cut, definitive predictions or even scenarios will be provided for the many complex issues that have been analysed and discussed here. Despite this absence, a reflection of the nature of things, it is hoped that this book has provided some insights into the workings of the Telecoms Industry, one of the most important postwar industries in the global economy, and perhaps into the processes of economic change more generally.

---

[8] As noted at several points in this book, for several reasons barriers to entry into some telecoms services markets, notably the local access market, remained high although not for technological reasons. See the example of COLT in Chapter 7 for further discussion.

[9] It is important also to stress that the creation and use of knowledge are interdependent processes. Both network operators and their equipment suppliers are jointly closely involved in the creation and use of knowledge and it is through the interdependencies in these processes that innovation in the industry is driven.

[10] As noted in the Preface to this book, the importance of changing cognitive visions in the evolution of the Telecoms Industry has led me to reject the conventional economics paradigm based on rational choice and optimization for the purpose of analysing the processes of change in this industry.

APPENDIX

*Why Michael Armstrong, CEO of AT&T, announced on
25 October 2000 that the company would be broken
up into four separate companies*

In Chapter 3 the reorganization of AT&T is compared to those of BT and NTT. In Chapters 4 and 5 the reorganizations of Deutsche Telekom and France Telecom (FT) are analysed. Significantly, AT&T is the only one of these Big Five incumbent network operators to announce[1] that it will break itself up completely. BT comes closest to AT&T as a result of its decision (examined in Chapter 3) to completely separate only its mobile subsidiary, formerly BT Wireless which was later renamed mmO2. However, Deutsche Telekom, FT, and NTT remained as loosely integrated single companies.

Since AT&T is the outlying case, it is important to try and understand exactly why Michael Armstrong, AT&T's CEO, announced on 25 October 2000 the decision to break the company up into four separate companies. At the time of writing, there is no comprehensive research that answers this question. In the absence of this detail, the present appendix makes use of a detailed interview given by Armstrong to *Business Week* in which he explains his decision.

THE REASONS FOR AT&T'S BREAK-UP

In a frank interview with *Business Week*,[2] AT&T's CEO, C. Michael Armstrong, provided his own *ex post* explanation. In answer to *Business Week's* question, 'Why break up AT&T?', Armstrong explained: 'To answer that, first you have to understand that in 1997, 82% of [AT&T's] business was long distance and it was going to go away [because of the significant increase in long distance

---

[1] At the time of writing the breakup has not been fully implemented and there has been some speculation in the financial press that the breakup decision may even be reversed. There is, however, not yet any sign of this happening.

[2] *Business Week*, 5 February 2001, p. 59.

competition from new entrants such as WorldCom and Qwest causing sharp falls in telephony prices]. So to save AT&T from becoming an American memory, we invested in three networks—wireless, cable, and data'.

It was Armstrong's acquisition of the cable TV companies, TCI and MediaOne, for a sum of $105 billion, that constituted AT&T's boldest move into local access, including broadband Internet access. Further investments were made in wireless and optical fibre networks. This gave AT&T substantial cable, wireless, and data (optical fibre) networks.

At the same time, Armstrong was busy developing AT&T's future strategy. But what precisely was this strategy?

AT&T's Strategy

Here the story becomes controversial. According to Armstrong, a common misconception prevailed regarding the strategy that AT&T was pursuing. This alleged misconception is evident in the second question that Armstrong was asked by *Business Week*: 'Many people believe the restructuring [i.e. break up of AT&T] reverses AT&T's "one-stop shop" strategy. Has it changed?'.

Armstrong's answer was that there was a failure to understand AT&T's strategy:

What some people thought they heard was that we would put your cable bill on your telephone bill, or put your telephone bill on your cable bill, and they called that bundling. I guess you'd call that cross-network bundling. That kind of bundling doesn't keep customers or attract customers to any degree. We never put a big emphasis on bundling all kinds of services, only those that travel over *the same network*. This confusion really frustrates me (emphasis added).

In fact, Armstrong insisted in the interview, 'The strategy was about bundling services that travel over *the same network*. For example, on our wireless network, we bundled a local wireless call, a roaming charge, and a long distance call—and charged a flat rate for it. That redefined the whole industry' (emphasis added).

Interestingly, in the same interview Armstrong also answers those critics who argue that like IBM, AT&T should have 'stuck to its knitting' and focused on its core business:

I thought it was really fascinating to read some of the things criticizing me. For example, if only I would just do like [IBM's] Lou Gerstner and focus on my core business, then AT&T could be much more successful. But the market did not understand that technology and deregulation are taking the middle of the [long distance] phone call away. It cannot exist by itself. The alternative today, had we not done what we did, was extinction.

But all this poses one very important question: Why is it that 'the market did not understand?' According to most economists one of the most important characteristics of financial markets is precisely their ability to collect and analyse all the relevant information. If it is possible for 'the market to misunderstand' then there are extremely important questions raised regarding the ability of financial markets to efficiently allocate resources.

Asked by *Business Week* 'Why do you believe your message [regarding AT&T's strategy and structure] has been coolly received?' Armstrong pointed the finger at AT&T's competitors, particularly the Baby Bells, who want to frustrate the company's success and therefore have a vested interest in spreading misinformation:

I suspect that a lot of this is fueled by people who don't want to see AT&T succeed. You think the [Baby Bells] aren't afraid of bundling over cable? I can put high-speed data, local and long-distance telephone service, video on demand, and hundreds of channels of digital video all together in one connection. You don't think they don't think that might be a tough thing to compete with? Damn right they do.

AT&T's Structure

According to Armstrong, it was after these investments in cable, wireless, and data networks were made that, 'we took a step back and said: What structure will best serve this strategy? That's when we decided to restructure [i.e. break AT&T up into four separate companies] ... '.

But why was a broken up structure—that is four separate companies: AT&T Wireless, AT&T Broadband [the cable business], AT&T Business Services, and AT&T Consumer Services—best for AT&T to successfully implement its strategy of bundling services over the same network? Armstrong attempts to answer this question in the following quotation (that includes some of the quotations given in the last paragraphs):

By 2000, we had defined our strategy, made the investments, bundled communications services on each one of those networks, and begun scaling the business. Then we took a step back and said: What structure will best serve this strategy?

That's when we decided to restructure [i.e. break AT&T up into four companies] for three reasons: currency, shareholder value, and employee motivation. Shareholders had been very patient while we made those investments. How was the value of all those investments going to be delivered to them? If we waited, would these businesses have missed the opportunity to strengthen themselves for industry consolidation? Would they have been able to keep the employees that they should keep and attract? And would shareholders have the patience to wait much longer? I judged not. It was time for the currencies, the equities, and the shareholder value to come through. I couldn't make that happen any other way.

Critical Reflections on Armstrong's Argument

In order to get a broader perspective on Armstrong's argument it is worth summarizing his logic in step-wise fashion:

1. AT&T's core business was long-distance telephony. However, under the pressure of competition and falling prices, this business was losing its ability to sustain AT&T's revenue growth and profitability.
2. The correct response was to invest in three kinds of networks—cable, wireless, and data (optical fibre)—that would provide local access as well as long-distance services and bundle a variety of services, both old and new, over each of them. Bundling over a specific network was the core of Armstrong's strategy.
3. With this strategy in place, the next step was to devise an appropriate corporate structure that would allow the strategy to be successfully implemented. This structure necessitated the break-up of AT&T into four separate companies.

How adequate is this logic in Armstrong's argument? Step 1 is an empirical issue and, since there is substantial evidence to support it, is relatively uncontroversial. Step 2 is a strategic choice issue. While it is possible to argue that another response, or variation of Armstrong's strategic choice, would have been more appropriate, only time will provide vindication or refutation of this choice. However, it is the logical link between Step 2 and Step 3 that is most questionable. The reason, simply, is that it is by no means clear why Armstrong's strategy necessitates this particular corporate structure (i.e. breakup) and no other. It is perfectly possible to argue that Armstrong's strategy could be effectively implemented with AT&T's four businesses remaining either a loosely integrated part of the same company (like Deutsche Telekom and France Telecom), or being organized as semi-autonomous businesses under the ultimate control of a holding company (like NTT). Accordingly, it is by no means obvious that Armstrong's strategy requires this particular structure and no other.

But there is a further question that has relevance for this issue. Are there 'synergies' or 'interdependencies' between AT&T's four businesses that provide benefits that would be lost if the company were broken up into four separate companies? If the answer is 'no', then the argument for breakup becomes stronger, particularly if there are important costs of integration, that is, costs of holding the four companies together (e.g. in terms of greater hierarchical complexity, slower decision-making, etc).

This question is dealt with in the special feature *Business Week* article accompanying the Armstrong interview. In this article it is stated that 'It was in November 1999, that [Armstrong] first considered busting up the company' (p. 56). The thought process began during discussions that Armstrong had with his friend Charles H. Noski, at the time president of Hughes Electronics Corp. In search of a CFO who would carry weight with Wall Street, Armstrong was trying to persuade Noski to take the job (which Noski did from January 2000). During these discussions, according to Noski, 'We challenged the theory that everything needed to be under one roof' (p. 56). At Hughes both Noski and Armstrong (then a Hughes employee) had created a tracking stock in order to separate in stock market valuation terms Hughes from its parent General Motors. The costs and benefits of integration versus separation was therefore a well-traversed issue for the two men.

According to the *Business Week* article,

Armstrong and Noski...had been researching how much the different AT&T businesses relied on each other. The answer? Not much. They estimated that less than 15% of AT&T's revenues came from cross-selling between two business units. That approach 'is not anything that either keeps customers or gets customers', says Armstrong. Over the summer [of 2000], Armstrong worked out the details of the four-way breakup. By the time the company's two-and-a-half day management retreat began on September 23 [2000] at AT&T headquarters in Basking Ridge, New Jersey, the restructuring was a working plan (p. 57).

How strongly does evidence that 'less than 15% of AT&T's revenues came from cross-selling between two business units' support Armstrong's argument that his strategy necessitated a break up structure? The answer must be that this evidence does not provide very strong support. First, 15 per cent of AT&T's revenues is not an insignificant amount—$10.1 billion on a total revenue of $67.1 billion. Indeed, the 2000 revenue of AT&T Wireless was $10.3 billion while that of AT&T Broadband (essentially the company's cable business bought from TCI and MediaOne) was $8.3 billion.

Second, there are other forms of 'synergy', or benefit from interdependence between businesses, apart from 'cross-selling between two business units'. For example, cross business information sharing or coordination may not show up in 'cross-selling' but may nonetheless provide important benefits from the integration of various businesses within the same overall company. An example from France Telecom's case is the many forms of cooperation that exist between its Wanadoo and New Orange subsidiaries in the areas of the Internet and mobile communications (see Chapter 5). While it is true that a fuller investigation of this issue would also have to take account of the possibly significant costs of integration of AT&T's four businesses (some of which have already been mentioned), it seems reasonable to conclude that the logical link holding together Steps 2 and 3 in Armstrong's argument is not nearly as tight as he appears to believe.

However, if this conclusion is correct it raises a key question: Why then did Armstrong make the decision to break up AT&T if it is not entirely clear that this was necessary to successfully implement his strategy? In order to answer this question we must go back to Armstrong's interview quoted above when he stated clearly that 'we decided to restructure (i.e. breakup) for three reasons: currency, shareholder value, and employee motivation'.

To begin with, it is necessary to explain what Armstrong means by these three explanatory variables. What does he mean by 'currency'? The answer is given elsewhere in the *Business Week* article where in an inset titled 'Grading the Chief' it is stated that one of Armstrong's achievements is that he 'used his currency. He made his cable buys with AT&T stock before the decline in the long-distance business destroyed the value of his currency' (p. 58). 'Currency', therefore, refers to the value of AT&T's shares which gives it purchasing power over other company's assets. What then is meant by 'shareowner value'? The answer must be that it refers to the return that a shareholder gets from the capital appreciation that occurs when the share price rises together with any dividend that is declared. Since the former usually outweighs the latter, it is clear that 'shareowner value' and 'currency' are very closely related.

What about 'employee motivation'? This third variable explaining the decision to break up AT&T refers to the attempt, through breakup, to more closely align performance and individual reward. A major mechanism for achieving this alignment is the granting of stock options. The problem with a consolidated, integrated company that does not report results for individual businesses and only has one share price is that employees may see very little relationship between their effort, which results in good performance in part of the company, and appreciation in the company's share price. Specifically in AT&T's case, employees in the wireless, broadband, and business companies may find that, due to falling revenue and profits in the consumer company, negatively affected by falling long-distance tariffs, AT&T's share price does not perform well despite their own successful efforts. Of course, it is just this that happened to AT&T's employees during the course of 2000. Significantly, the *Business Week* article also stated that 'In a development not yet made public, Armstrong persuaded his board to issue a special batch of new [stock] options for as many as 56,000 eligible employees' (p. 55).

Clearly, therefore, the solution to the problem of 'employee motivation' was seen by Armstrong as being very closely related to the company's share price. Accordingly, there is a significant degree of overlap between Armstrong's three reasons for breakup—currency, shareowner value, and employee motivation— in so far as all three are highly dependent on AT&T's share price performance.

It may be concluded, therefore, according to Armstrong's own account of why he decided to break up AT&T, that the main reason had to do with an attempt to restore the company's flagging share price. The argument that only through breakup could AT&T successfully implement its strategy is, as we have shown, not particularly convincing even though it might be used by Armstrong as an additional weapon in the fight to rationalize his decision to break the company up to financial analysts and investors. The attempt to raise AT&T's share price, therefore, was the main driving force behind Armstrong's breakup plan.

The financial markets, however, were less convinced that breakup would make a significant improvement in AT&T's revenue and earnings. The day the breakup was announced, AT&T's share price declined by 13 per cent and the shares ended 66 per cent down over the year 2000 (though this also reflected the general bear market in technology and telecoms shares in the latter half of 2000). *The New York Times* commented that 'If only AT&T worked as hard at telecommunications engineering as it does at financial engineering, maybe investors would treat it with more respect'. According to the *Business Week* article, 'Armstrong says it [i.e. the announcement of breakup] was one of the lowest points in his career: "It was a huge disappointment"' (p. 57).

BIBLIOGRAPHY

Abbate, J. (1999). *Inventing the Internet*. Cambridge, MA: MIT Press.

Aksoy, A. (1992). 'Mapping the Information Business: Integration for Flexibility', in K. Robins (ed.), *Understanding Information Business, Technology and Geography*. London: Belhaven Press, pp. 43–62.

Ancori, B., Bureth, A., and Cohendet, P. (2000). 'The Economics of Knowledge: The Debate about Codification and Tacit Knowledge'. *Industrial and Corporate Change*, 9(2): 255–87.

Antonelli, C. (1999). *The Microdynamics of Technological Change*. London: Routledge.

—— (2001). *The Microeconomics of Technological Systems*. Oxford: Oxford University Press.

Armstrong, M. (1997). 'Competition in Telecommunications'. *Oxford Review of Economic Policy*, 13(1): 64–82.

—— Cowan, S. and Vickers, J. (1994). *Regulatory Reform, Economic Analysis and British Experience*. Cambridge, MA: MIT Press.

—— Doyle, C. and Vickers, J. (1996). 'The Access Pricing Problem: A Synthesis'. *Journal of Industrial Economics*, 44: 131–50.

Arrow, K. (1974). *The Limits of Organization*. New York: W. W. Norton.

—— Carlton, D. W., and Sider, H. S. (1995). 'The Competitive Effects of Line-of-Business Restrictions in Telecommunications'. *Managerial and Decision Economics*, 16.

Arthur, W. B. (1988). 'Competing Technologies: An Overview', in G. Dosi, C. Freeman, R. Nelson, G. Silverberg, and L. Soete (eds), *Technical Change and Economic Theory*. London: Pinter.

Asanuma, B. (1989). 'Manufacturer–Supplier Relationships in Japan and the Concept of Relation-specific Skill'. *Journal of the Japanese and International Economies*, 3: 1–30.

Audretsch, D. B. (1977). 'Technological Regimes, Industrial Demography and the Evolution of Industrial Structures'. *Industrial and Corporate Change*, 6(1): 49–82.

Bailey, J., *et al.* (1995). 'The Economics of Advanced Services in an Open Communications Infrastructure: Transaction Costs, Production Costs, and Network Externalities'. *Information Infrastructure and Policy*, 4: 255–78.

Baldwin, C. Y. and Clark, K. B. (2000). *Design Rules, Volume 1. The Power of Modularity*. Cambridge, MA: MIT Press.

Bar, F. (1991). 'Network Flexibility: A New Challenge for Telecom Policy'. *Communications and Strategies*, 2: 113–23.

Batty, M. and Barr, B. (1994). 'The Electronic Frontier: Exploring and Mapping Cyberspace'. *Futures*, 26(7): 699–712.

Berners-Lee, T., Cailliau, R., Luotonen, A., Frystyk Nielsen, H., and Secret, A. (1994). 'The World-Wide Web'. *Communications of the ACM*, 37(8): 76–82.

Bertolotti, M. (1983). *Masers and Lasers: An Historical Approach*. Bristol: Adam Hilger.

Bijker, W. E., Hughes, T. P., and Pinch, T. (eds) (1987). *The Social Construction of Technological Systems*. Cambridge, MA: MIT Press.

Bohlin, E. and Levin, S. (eds) (1998). *Telecommunications Transformation, Technology, Strategy and Policy*. Amsterdam: IOS Press.

Bourreau, M. and Dogan, P. (2001). 'Regulation and Innovation in the Telecommunications Industry'. *Telecommunications Policy*, 25(3): 167–84.

Brock, G. W. (1994). *Telecommunication Policy for the Information Age: From Monopoly to Competition*. Cambridge, MA: Harvard University Press.

Brooks, J. (1975). *Telephone: The First Hundred Years*. New York: Harper & Row.

Brousseau, E. and Quélin, B. (1992). 'Users' Knowledge as a Specific Asset: The Case of Value Added Services'. *International Journal of Information Technology*, 7: 233–43.

——and Quélin, B. (1996). 'Asset Specificity and Organizational Arrangements: The Case of the New Telecommunications Services Market'. *Industrial and Corporate Change*, 5(4): 1205–30.

Campbell-Kelly, M. (1988). 'Data Communications at the National Physical Laboratory (1965–1975)'. *Annals of the History of Computing*, 9(3): 221–47.

Carpentier, M., Farnoux-Toporkoff, S., and Garric, C. (1992). *Telecommunications in Transition*. Chichester: John Wiley & Sons.

Cave, M. (1997). 'The Evolution of Telecommunications Regulation in the UK'. *European Economic Review*, 41: 691–9.

——and Williamson, P. (1996). 'Entry, Competition and Regulation in UK Telecommunications'. *Oxford Review of Economic Policy*, 12(4): 100–21.

Cerf, V. (1993). 'How the Internet Came to Be', in B. Aboba (ed.), *The Online User's Encyclopedia*. Addison-Wesley.

——(1996). 'Is there a Future for the Net? David Pitchford finds out from the Man who Invented it, Vinton Cerf'. *Internet*, 19 June: 75.

Chaffee, C. D. (1988). *The Rewiring of America: The Fiber Optics Revolution*. Boston, MA: Academic Press.

Chandler, A. D. (1969). *Strategy and Structure: Chapters in the History of the Industrial Enterprise*. Cambridge, MA: MIT Press.

——(1977). *The Visible Hand: The Managerial Revolution in American Business*. Cambridge, MA: The Belknap Press.

——(1984). 'The Emergence of Managerial Capitalism'. *Business History Review*, 58: 473–503.

——(1990). *Scale and Scope: The Dynamics of Industrial Capitalism*. Cambridge, MA: The Belknap Press.

——and Salsbury, S. (1971). *Pierre S. du Pont and the Making of the Modern Corporation*. New York: Harper & Row.

Chapuis, R. J. and Joel, A. E. (1990). 100 *Years of Telephone Switching*, ii. *Electronics, Computers and Telephone Switching: A Book of Technological History*, 1960–1985. Amsterdam: North-Holland.

Cohendet, P. and Steinmueller, W. E. (2000). 'The Codification of Knowledge: A Conceptual and Empirical Exploration'. *Industrial and Corporate Change*, 9(2): 195–209.

Computer Science and Telecommunications Board, National Research Council. (1994). *Realizing the Information Future: The Internet and Beyond.* Washington, DC: National Academy Press.

—— (1996). *The Unpredictable Certainty: Information Infrastructure Through 2000.* Washington, DC: National Academy Press.

Coopersmith, J. (1993). 'Facsimile's False Starts'. *IEEE Spectrum*, Feb.: 46–9.

Cowan, R., David, P. A., and Foray, D. (2000). 'The Explicit Economics of Knowledge Codification and Tacitness'. *Industrial and Corporate Change*, 9(2): 211–53.

Crandall, R. W. and Waverman, L. (1996). *Talk is Cheap: Declining Costs, New Competition, and Regulatory Reform in Telecommunications.* Brookings Institute.

David, P. A. (1986). 'Understanding the Economics of QWERTY: The Necessity of History', in W. N. Parker (ed.), *Economic History and the Modern Economist.* Oxford: Basil Blackwell.

—— (1988). 'Path-dependence: Putting the Past into the Future of Economics'. *Technical Report* 533, Economics Series. Institute for Mathematical Studies in the Social Sciences, Stanford University, Palo Alto, California.

—— and Steinmueller, W. E. (1994). 'Economics of Compatibility Standards and Competition in Telecommunications Networks'. *Information Economics and Policy*, 6: 217–42.

Department of Trade and Industry. (1991). *Competition and Choice: Telecommunications Policy for the 1990s.* London: HMSO.

Dertouzos, M. L., Lester, R. K., and Solow, R. M. (1989). *Made in America: Regaining the Productivity Edge.* Cambridge, MA: MIT Press.

De Sola Pool, I. and Noam, E. M. (eds). (1990). *Technologies Without Boundaries: On Telecommunications in a Global Age.* Cambridge, MA: Harvard University Press.

Dixit, A. (1979). 'A Model of Duopoly Suggesting a Theory of Entry Barriers'. *Bell Journal of Economics*, 10: 20–32.

Dosi, G. (1982). 'Technological Paradigms and Technological Trajectories: A Suggested Interpretation of the Determinants and Directions of Technical Change'. *Research Policy*, 11: 147–63.

—— (1990). 'Finance, Innovation, and Industrial Change'. *Journal of Economic Behavior and Organization*, 13: 299–319.

—— Malerba, F., Marsili, O., and Orsenigo, L. (1997). 'Industrial Structures and Dynamics: Evidence, Interpretations and Puzzles'. *Industrial and Corporate Change*, 6(1): 3–24.

—— Teece, D. J., and Chytry, J. (eds). (1998). *Technology Organisation and Competitiveness: Perspectives on Industrial and Corporate Change.* Oxford: Oxford University Press.

Dretske, F. I. (1982). *Knowledge and the Flow of Information.* Cambridge, MA: MIT Press.

Duke, D. A. (1983). 'A History of Optical Communications'. *Special Report.* Telecommunication Products Department, Corning Glass Works, Corning, New York.

Einhorn, M. A. (1996). 'Internet Voice, "Cyberpass", Competitive Efficiency'. *Industrial and Corporate Change*, 5(4): 1067–78.

Flamm, K. (1987). *Targeting the Computer: Government Support and International Competition.* Washington, DC: The Brookings Institution.

—— (1988). *Creating the Computer: Government, Industry, and High Technology.* Washington, DC: The Brookings Institution.

Fransman, M. (1992a). 'Controlled Competition in the Japanese Telecommunications Equipment Industry: The Case of Central Office Switches', in C. Antonelli (ed.), *The Economics of Information Networks.* Amsterdam: North-Holland.

—— (1992b). 'Japanese Failure in a High-tech Industry? The Case of Central Office Telecommunications Switches'. *Telecommunications Policy*, 16(3): 259–76.

Fransman, M. (1993). *The Market and Beyond: Information Technology in Japan*. Cambridge: Cambridge University Press.

—— (1994*a*). 'AT&T, BT and NTT: A Comparison of Vision, Strategy and Competence'. *Telecommunications Policy*, 18(2): 137–53.

—— (1994*b*). 'AT&T, BT and NTT: The Role of R&D'. *Telecommunications Policy*, 18(4): 295–305.

—— (1994*c*). 'Knowledge Segmentation-integration in Theory and in Japanese Companies', in O. Granstrand (ed.), *Economics of Technology*. Amsterdam: North-Holland, pp. 170–5.

—— (1994*d*). 'Different Folks, Different Strokes: How IBM, AT&T and NEC Segment to Compete'. *Business Strategy Review*, 5(3): 1–20.

—— (1994*e*). 'Information, Knowledge, Vision and Theories of the Firm'. *Industrial and Corporate Change*, 3(2): 1–45.

—— (1995). *Japan's Computer and Communications Industry: The Evolution of Industrial Giants and Global Competitiveness*. Oxford: Oxford University Press.

—— (1998). 'Information, Knowledge, Vision and Theories of the Firm', in G. Dosi, D. J. Teece, and J. Chytry (eds), *Technology, Organization, and Competitiveness: Perspectives on Industrial and Corporate Change*. New York: Oxford University Press.

—— (1999*a*). *Visions of Innovation: The Firm and Japan*. Oxford: Oxford University Press.

—— (1999*b*). 'Where are the Japanese? Japanese Information and Communications Firms in an Internetworked World'. *Telecommunications Policy*, 23: 317–33.

—— (2000*a*). 'Convergence, the Internet and Multimedia: Implications for the Evolution of Industries and Technologies', in E. Bohlin, K. Brodin, A. Lundgren, and B. Thorngren (eds), *Convergence in Communications and Beyond*. Amsterdam: Elsevier.

—— (2000*b*). 'AT&T, BT and NTT: Super-Adapters or Dinosaurs caught in a time of Climate Change?' Mimeo.

—— (2002). 'Vodafone's Rise as a Global Mobile Operator'. Mimeo.

—— (forthcoming). 'Evolution of the Telecoms Industry into the Internet Age', in *International Handbook on the Economics of Telecommunications*. Edward Elgar.

Fudenberg, D. and Tirole, J. (1983). 'Capital as a Commitment: Strategic Investment to Deter Mobility'. *Journal of Economic Theory*, 31: 227–56.

Fujitsu. (1977). *The History of Fujitsu*. Tokyo: Fujitsu (in Japanese).

Garrette, B. and Quélin, B. (1994). 'An Empirical Analysis of Hybrid Forms of Governance Structures: The Case of Telecommunication Equipment Industry'. *Research Policy*, 23: 395–412.

Gilbert, R. (1987). 'Mobility Barriers and the Value of Incumbency', in R. Schmalensee and R. D. Willig (eds), *The Handbook of Industrial Organization*. Amsterdam: North-Holland.

Gong, J. and Srinagesh, P. (1996). 'Network Competition and Industry Structure'. *Industrial and Corporate Change*, 5(4): 1231–41.

—— and Srinagesh, P. (1997). 'The Economics of Layered Networks', in L. McKnight and J. Bailey (eds), *Internet Economics*. Cambridge, MA: MIT Press.

Grossman, S. and Hart, O. (1986). 'The Costs and Benefits of Ownership: A Theory of Vertical and Lateral Integration'. *Journal of Political Economy*, 94: 691–719.

Grove, A. S. (1996). *Only the Paranoid Survive*. New York: Doubleday.

Grubman, J. G., *et al.* (1996). *Global Telecommunications Jigsaw Puzzle: Increasing Demand for International Network Services Drives Industry Consolidation*. New York: Salomon Brothers, 16 Dec.

—— (1997). *MFS WorldCom—The First Phone Company for the New Millenium*. New York: Salomon Brothers, 2 Jan.

—— and McMahon, S. (1998*a*). *WorldCom, Inc. Combination with MCI Creates the Only Legitimate Telecom Large-cap Growth Stock*. New York: Salomon Smith Barney, 9 Apr.

—— *et al.* (1998*b*). *WorldCom: Spectacular 1998 Results Highlight Diversity of Revenues and Reach of Assets*. New York: Salomon Smith Barney, 4 May.

Hafner, K. and Lyon, M. (1996). *Where Wizards Stay Up Late: The Origins of the Internet*. Simon & Schuster.

Harrington, A., *et al.* (1997). *European Telecommunications: Liberalisation—A Brave New World*. London: Salomon Brothers, 21 Mar.

—— *et al.* (1998*a*). *Energis. Poles Apart*. London: Salomon Smith Barney, 15 Jan.

—— and Sloane, R. (1998*b*). *COLT: A Premier New Entrant Opportunity*. London: Salomon Smith Barney, 22 Jan.

Harris, R. (1966). 'Electronic Telephone Exchanges: An Introductory Review of Development'. *POEEJ*, 59(3): 211–19.

—— and Martin, J. (1981). 'The Evolution of Switching Systems Architecture'. *POEEJ*, 74: 187–93.

Harvard Business School. (1993). *DDI Corporation*. Case Study 9-393-048. Boston, MA: Harvard Business School.

Hawkins, R. W., Mansell, R. E., and Skea, J. (eds) (1995). *Standards, Innovation and Competitiveness: The Politics and Economics of Standards in Natural and Technical Environments*. Aldershot: Edward Elgar.

—— Mansell, R. E., and Steinmueller, W. E. (1997). *Green Paper: Mapping and Measuring the Information, Technology, Electronics and Communications Sector in the United Kingdom*. London: Department of Trade and Industry.

Henderson, R. M. and Clark, K. B. (1990). 'Architectural Innovation: The Reconfiguration of Existing Product Technologies and the Failure of Established Firms'. *Administrative Science Quarterly*, 35: 9–30.

Houghton, J. W. (1999). 'Mapping Information Industries and Markets'. *Telecommunications Policy*, 23(10): 689–99.

Huber, P. W. (1987). *The Geodesic Network: Report on Competition in the Telephone Industry*. Washington, DC: US Department of Justice.

Hughes, T. P. (1984) *Networks of Power: Electrification in Western Society, 1880–1930*. Baltimore, Maryland: Johns Hopkins University Press.

Inose, H. (1979). *An Introduction to Digital Integrated Communications Systems*. Tokyo: University of Tokyo Press.

—— Nishikawa, T., and Uenohara, M. (1982). 'Cooperation between Universities and Industries in Basic and Applied Science', in A. Gerstenfeld (ed.), *Science Policy Perspectives: USA-Japan*. Tokyo: Academic Press.

Inoue, T. (1990). *NTT: Kyoso to bunkatsu ni chokumensuru jokoka jidai no kyojin* (NTT: A Giant in the Information-oriented Age Facing Competition and Partition). Tokyo: Otsukishoten.

Johnson, C. (1989). 'MITI, MPT, and the Telecom Wars: How Japan Makes Policy for High Technology', in C. Johnson, L. D. Tyson, and J. Zysman (eds), *Politics and Productivity: How Japan's Development Strategy Works*. New York: Ballinger.

Kaneko, H. (1990). *NTT no mirai aratana jigyotenkai to bunkatsu no shogeki* (The Future of NTT). Tokyo: Toyokeizai shinpo.

Kano, S. (2000). 'Technical Innovations, Standardization and Regional Comparison—A Case Study in Mobile Communications'. *Telecommunications Policy*, 24: 305–21.

Kao, C. and Hockham, G. A. (1966). 'Dielectric-fiber Surface Waveguides for Optical Frequencies'. *Proceedings of the IEEE*, 113(7): 1151–8.

Katz, M. L. (1989). 'Vertical Contractual Relationships', in R. Schmalensee and R. D. Willig (eds), *The Handbook of Industrial Organization*. Amsterdam: North-Holland, pp. 655–721.

—— (1996). 'Remarks on the Economic Implications of Convergence'. *Industrial and Corporate Change*, 5(4): 1079–95.

—— and Shapiro, C. (1985). 'Network Externalities, Competition, and Compatibility'. *American Economic Review*, 75: 424–40.

—— and Shapiro, C. (1994). 'Systems Competition and Network Effects. *Journal of Economic Perspectives*, 8: 93–115.

Kavassalis, P., *et al.* (1996). 'The Internet: A Paradigmatic Rupture in Cumulative Telecom Evolution'. *Industrial and Corporate Change*, 5: 1097–126.

—— Lee, T., and Bailey, J. (1998). 'Sustaining a Vertically Disintegrated Network Through a Bearer Service Market', in E. Bohlin and S. Levin (eds), *Telecommunications Transformation, Technology, Strategy and Policy*. Amsterdam: IOS Press.

—— and Lehr, W. (1998). 'Forces for Integration and Disintegration in the Internet'. *Communications and Strategies*, 30.

—— and Solomon, R. (1997). 'Mr. Schumpeter on the Telephone: Patterns of Technical Change in the Telecommunications Industry Before and After the Internet'. *Communications and Strategies*, 26: 371–408.

Kennedy, P., *et al.* (2001). *COLT Telecom Group*. London: Morgan Stanley Dean Witter, 18 Sept.

Kitahara, Y. (1983). *Information Network System: Telecommunications in the Twenty-first Century*. London: Heinemann.

Klepper, S. (1997). 'Industry Life Cycles'. *Industrial and Corporate Change*, 6: 145–83.

—— and Simons, K. L. (1997). 'Technological Extinctions of Industrial Firms: An Inquiry into Their Nature and Causes'. *Industrial and Corporate Change*, 6: 379–460.

Knight, F. (1921). *Risk, Uncertainty and Profit*. Boston, MA: Houghton Mifflin.

Kobayashi, K. (1984). 'The Past, Present and Future of Telecommunications in Japan'. *IEEE Communications Magazine*, 22(5): 97.

—— (1989). *Rising to the Challenge: The Autobiography of Koji Kobayashi*. Tokyo: Harcourt Brace Jovanovich.

—— (1991). *The Rise of NEC: How the World's Greatest C&C Company is Managed*. Oxford: Blackwell.

Kobayashi, T. (1986). *Fortune Favors the Brave. Fujitsu: Thirty Years in Computers*. Tokyo: Toyo Keizai Shinposha.

Lachmann, L. (1978). *Capital and its Structure*. Kansas City, Missouri: Sheed Andrews and McMeel.

Laffont, J. -J. and Tirole, J. (1994). *Creating Competition through Interconnection: Theory and Practice*. Toulouse: IDEI.

—— Rey, P., and Tirole, J. (1996). 'Network Competition: I. Overview and Non-discriminatory Pricing'. *RAND Journal of Economics*.

—— Rey, P., and Tirole, J. (1997). 'Competition Between Telecommunications Operators'. *European Economic Review*, 41: 701–11.

Langlois, R. N. and Robertson, P. L. (1995). *Firms, Markets and Economic Change: A Dynamic Theory of Business Institutions.* London: Routledge.

Leiner, B. M., Cerf, V. G., Clark, D. D., Kahn, R. E., Kleinrock, L., Lynch, D. C., Postel, J., Roberts, L. G., and Wolff, S. (1997). 'A Brief History of the Internet'. Web page, http://www.isoc.org/internet-history

Loasby, B. (2000). 'Decision Premises, Decision Cycles and Decomposition'. *Industrial and Corporate Change,* 9(4): 709–32.

——(2001*a*). 'Cognition, Imagination and Institutions in Demand Creation'. *Journal of Evolutionary Economics,* 11(1): 7–22.

——(2001*b*). 'Time, Knowledge and Evolutionary Dynamics: Why Connections Matter. *Journal of Evolutionary Economics,* 11(4): 393–412.

Machlup, F. (1962). *The Production and Distribution of Knowledge in the United States.* Princeton, NJ: Princeton University Press.

McKelvey, M. (forthcoming). 'Global Firms, Small Countries: Ericsson, Nokia and Wireless Telecommunications'. *International Journal of Technology Management.*

——Texier, F., and Alm, H. (1998). 'The Dynamics of High Tech Industry: Swedish Firms Developing Mobile Telecommunication Systems'. Mimeo.

McKnight, L. and Bailey, J. (eds) (1997). *Internet Economics.* Cambridge, MA: MIT Press.

McLaughlin, J. F. and Antonoff, A. L. (1986). 'Mapping the Information Business'. Harvard University Program on Information Resources Policy P-86-9. Cambridge, MA, Sept.

Maiman, T. H. (1987). *The Laser: Its Origin, Applications and Future.* Japan Prize 1987 Official Brochure, pp. 12–17.

Malerba, F. and Orsenigo, L. (1990). 'Technological Regimes and Patterns of Innovation: A Theoretical and Empirical Investigation of the Italian Case', in A. Heertje (ed.), *Evolving Industries and Market Structures.* Ann Arbor, Michigan: University of Michigan Press.

——and Orsenigo, L. (1993). 'Technological Regimes and Firm Behaviour'. *Industrial and Corporate Change,* 2: 45–71.

——(1995). 'Schumpeterian Patterns of Innovation'. *Cambridge Journal of Economics,* 19(1): 47–66.

——and Orsenigo, L. (2000). 'Knowledge, Innovative Activities and Industrial Evolution'. *Industrial and Corporate Change,* 9(2): 289–314.

Malkiel, B. (1987). 'Efficient Market Hypothesis', in J. Eatwell, M. Milgate, and P. Newman (eds), *The New Palgrave: A Dictionary of Economics.* London: Macmillan, pp. 120–3.

Maney, K. (1995). *Megamedia Shakeout: The Inside Story of the Leaders and the Losers in the Exploding Communications Industry.* New York: John Wiley & Sons.

Mansell, R. (1994). *The New Telecommunications: A Political Economy of Network Evolution.* London: Sage Publications.

March, J. G. (1994). *A Primer on Decision Making: How Decisions Happen.* New York: Free Press.

Marshall, A. (1969). *Principles of Economics.* London: Macmillan.

Metcalfe, J. S. (1995). 'Technology Systems and Technology Policy in an Evolutionary Framework'. *Cambridge Journal of Economics,* 19(1): 25–46.

Meurling, J. (1985). *A Switch in Time.* Chicago, IL: Telephony Publishing Co.

——and Jeans, R. (1994). *The Mobile Phone Book: The Invention of the Mobile Phone Industry.* London: Communications Week International.

Millman, S. (ed.) (1983). *A History of Engineering and Science in the Bell System: Physical Sciences* (1925–1980). Murray Hill, NJ: AT&T Bell Laboratories.

Millman, S. (ed.) (1984). *A History of Engineering and Science in the Bell System: Communications Sciences* (1925–1980). Murray Hill, NJ: AT&T Bell Laboratories.

Miyazaki, K. (1995). *Building Competences in the Firm: Lessons from Japanese and European Optoelectronics*. London: Macmillan.

Mokyr, J. (1995). 'Comment on "The Boundaries of the US Firm in R&D" by David C. Mowery', in N. R. Lamoreaux and D. M. G. Raff (eds), *Coordination and Information*. Chicago: The University of Chicago Press.

Morton, J. A. (1971). *Organizing for Innovation: A Systems Approach to Technical Management*. New York: McGraw-Hill.

Mowery, D. C. (1995). 'The Boundaries of the US Firm in R&D', in N. R. Lamoreaux and D. M. G. Raff (eds), *Coordination and Information*. Chicago: The University of Chicago Press.

——and Nelson, R. R. (eds) (1999). *The Sources of Industrial Leadership*. Cambridge: Cambridge University Press.

Nagai, S. (1990). 'On the Competition of Telecommunications under Regulation in Japan'. *Journal of International Economic Studies*, 4: 15–32.

Nelson, R. R. (ed.) (1993). *National Innovation Systems: A Comparative Analysis*. New York: Oxford University Press.

——(1982). *An Evolutionary Theory of Economic Change*. Cambridge, MA: The Belknap Press.

——(1998). 'The Co-evolution of Technology Industrial Structure and Supporting Institutions', in G. Dosi, D. J. Teece, and J. Chytry (eds), *Technology, Organization and Competitiveness: Perspectives on Industrial and Corporate Change*. Oxford: Oxford University Press.

——and Winter, S. G. (1974). 'Neoclassical vs Evolutionary Theories of Economic Growth: Critique and Prospectus'. *Economic Journal*, 84: 886–905.

——and Winter, S. G. (1978). 'Forces Generating, and Limiting Concentration under Schumpeterian Competition'. *Bell Journal of Economics*, 9: 524–48.

——and Winter, S. G. (1982). *An Evolutionary Theory of Economic Change*. Cambridge, MA: Harvard University Press.

Noam, E. M. (1994). 'Beyond Liberalization: From the Network of Networks to the System of Systems'. *Telecommunications Policy*, 18: 286–94.

——(1995). *The Impending Doom of Common Carriage*. Aspen Institute.

——and Pogorel, G. (eds) (1994). *Asymmetric Deregulation: The Dynamics of Telecommunications Policy in Europe and the United States*. Ablex Publishing Corporation.

——and Wolfson, A. J. (eds) (1997). *Globalism and Localism in Telecommunications*. Elsevier Science Ltd.

Norberg, A. L. and O'Neill, J. E. (1996). *Transforming Computer Technology: Information Processing for the Pentagon*, 1962–1986. Baltimore, Maryland: Johns Hopkins University Press.

Patel, C. K. N. (1987). 'Lasers in Communications and Information Processing', in J. H. Ausubel and H. D. Langford (eds), *Lasers: Invention to Application*. Washington, DC: National Academy Press.

Penrose, E. T. (1955). 'Limits to the Growth and Size of Firms'. *American Economic Review*, 45(2): 531–43.

——(1959). *The Theory of the Growth of the Firm*. Oxford: Basil Blackwell.

Penzias, A. A. (1987). *The Grace A. Tanner Lecture: In Human Values (Computer-enhanced Human Beings).* Tanner Center for Human Values.

——(1989). *Ideas and Information: Managing in a High-tech World.* New York: W.W. Norton & Co.

——(1995). *Digital Harmony: Business, Technology and Life After Paperwork.* HarperCollins.

Rao, P. M. (1996). 'R&D and Innovation in US Telecommunications: Recent Structural Changes and Their Implications', in *Proceedings of the International Telecommunications Society Eleventh Biennial Conference.*

——(1998). 'The Changing Structure of Corporate R&D in US Telecommunications: Is Software Taking the Helm?' *International Telecommunications Society Twelfth Biennial Conference.* Mimeo.

Reich, L. S. (1985). *The Making of American Industrial Research: Science and Business at GE and Bell,* 1876–1926. Cambridge: Cambridge University Press.

Richardson, G. B. (1972). 'The Organisation of Industry'. *The Economic Journal,* 82(327): 883–96.

Roberts, L. G. (1978). 'The Evolution of Packet Switching'. *Proceedings of the IEEE,* 66(11): 1307–13.

——(1988). 'The ARPANET and Computer Networks', in A. Goldberg (ed.), *A History of Personal Workstations.* ACM Press.

Rosenberg, N. (1982). *Inside the Black Box: Technology and Economics.* Cambridge: Cambridge University Press.

——(1994). *Exploring the Black Box: Technology, Economics and History.* Cambridge: Cambridge University Press.

Salomon Smith Barney. (Various). *European Telecom Monthly.* London.

Salus, P. H. (1995). *Casting the NET: From ARPANET to Internet and Beyond.* Addison-Wesley.

Sandbach, J. (2001). 'Levering Open the Local Loop: Shaping BT for Broadband Competition'. *info,* 3(3): 195–202.

Schumpeter, J. A. (1943). *Capitalism, Socialism and Democracy.* London: Unwin.

——(1954). *History of Economic Analysis.* New York: Oxford University Press.

——(1961). *The Theory of Economic Development: An Inquiry into Profits, Capital, Credit, Interest and the Business Cycle.* New York: Oxford University Press.

——(1966). *Capitalism, Socialism and Democracy.* London: Unwin.

Shapiro, C. and Varian, H. R. (1999). *Information Rules: A Strategic Guide to the Network Economy.* Boston, MA: Harvard Business School Press.

Shiller, R. J. (2000, new Preface 2001). *Irrational Exuberance.* Princeton, NJ: Princeton University Press.

Shockley, W. (1950). *Electronics and Holes in Semiconductors, with Applications to Transistor Electronics.* New York: van Nostrand.

Simon, H. A. (1957). *Models of Man.* New York: John Wiley & Sons.

——Egidi, M., Marris, R., and Viale, R. (1992). *Economics, Bounded Rationality and the Cognitive Revolution.* Aldershot: Edward Elgar.

Sinclair, T., Grubman, J. B., and Dodd, R. (1999). *Viatel Inc. Total Transformation.* New York: Salomon Smith Barney, 6 Oct.

Stehmann, O. (1995). *Network Competition for European Telecommunications.* Oxford: Oxford University Press.

Steinmueller, W. E. (2000). 'Will New Information and Communication Technologies Improve the "Codification" of Knowledge?'. *Industrial and Corporate Change*, 9(2): 361–76.

Teece, D. J. (1993). 'The Dynamics of Industrial Capitalism: Perspectives on Alfred Chandler's *Scale and Scope* (1990)'. *Journal of Economic Literature*, 31.

—— (1995). 'Telecommunications in Transition: Unbundling, Reintegration and Competition'. *Michigan Telecommunications and Technology Law Review*, 4.

—— Rumelt, R., Dosi, G., and Winter, S. (1994). 'Understanding Corporate Coherence: Theory and Evidence'. *Journal of Economic Behavior and Organization*, 23: 1–30.

—— and Waverman, L. (eds) (1998). *Privatization, Deregulation and the Transition to Markets*. The Globalization of the World Economy Series. Aldershot: Edward Elgar.

Temin, P. and Galambos, L. (1987). *The Fall of the Bell System: A Study in Prices and Politics*. Cambridge: Cambridge University Press.

Townes, C. H. (1965). '1964 Nobel Lecture: Production of Coherent Radiation by Atoms and Molecules'. *IEEE Spectrum*, Aug.: 30–43.

Waverman, L. (1990). 'R&D and Preferred Supplier Relationships: The Growth of Northern Telecom'. Paper presented to the International Telecommunications Society (ITS) Conference, Venice. Mimeo.

—— and Sirel, E. (1997). 'European Telecommunications Markets on the Verge of Full Liberalization'. *Journal of Economic Perspectives*, 11(4): 113–26.

Weare, C. (1996). 'Interconnections: A Contractual Analysis of the Regulation of Bottleneck Telephone Monopolies'. *Industrial and Corporate Change*, 5(4): 963–92.

Winter, S. G. (1984). 'Schumpeterian Competition in Alternative Technological Regimes'. *Journal of Economic Behavior and Organization*, 5: 287–320.

Witt, U. (2000). 'Changing Cognitive Frames—Changing Organizational Forms: An Entrepreneurial Theory of Organizational Development'. *Industrial and Corporate Change*, 9(4): 733–56.

—— (2001). 'Economic Growth—What Happens on the Demand Side?'. *Journal of Evolutionary Economics*, 11(1): 1–6.

INDEX